Springer Theses

Recognizing Outstanding Ph.D. Research

Aims and Scope

The series "Springer Theses" brings together a selection of the very best Ph.D. theses from around the world and across the physical sciences. Nominated and endorsed by two recognized specialists, each published volume has been selected for its scientific excellence and the high impact of its contents for the pertinent field of research. For greater accessibility to non-specialists, the published versions include an extended introduction, as well as a foreword by the student's supervisor explaining the special relevance of the work for the field. As a whole, the series will provide a valuable resource both for newcomers to the research fields described, and for other scientists seeking detailed background information on special questions. Finally, it provides an accredited documentation of the valuable contributions made by today's younger generation of scientists.

Theses are accepted into the series by invited nomination only and must fulfill all of the following criteria

- They must be written in good English.
- The topic should fall within the confines of Chemistry, Physics, Earth Sciences, Engineering and related interdisciplinary fields such as Materials, Nanoscience, Chemical Engineering, Complex Systems and Biophysics.
- The work reported in the thesis must represent a significant scientific advance.
- If the thesis includes previously published material, permission to reproduce this must be gained from the respective copyright holder.
- They must have been examined and passed during the 12 months prior to nomination.
- Each thesis should include a foreword by the supervisor outlining the significance of its content.
- The theses should have a clearly defined structure including an introduction accessible to scientists not expert in that particular field.

More information about this series at http://www.springer.com/series/8790

Xiaoyi Liu

Nanomechanics of Graphene and Design of Graphene Composites

Doctoral Thesis accepted by
University of Science and Technology of China,
Hefei, China

 Springer

Author
Dr. Xiaoyi Liu
Department of Modern Mechanics
University of Science and Technology
of China
Hefei, China

Supervisor
Prof. Hengan Wu
University of Science
and Technology of China
Hefei, China

ISSN 2190-5053   ISSN 2190-5061   (electronic)
Springer Theses
ISBN 978-981-13-8702-9   ISBN 978-981-13-8703-6   (eBook)
https://doi.org/10.1007/978-981-13-8703-6

© Springer Nature Singapore Pte Ltd. 2019
This work is subject to copyright. All rights are reserved by the Publisher, whether the whole or part of the material is concerned, specifically the rights of translation, reprinting, reuse of illustrations, recitation, broadcasting, reproduction on microfilms or in any other physical way, and transmission or information storage and retrieval, electronic adaptation, computer software, or by similar or dissimilar methodology now known or hereafter developed.
The use of general descriptive names, registered names, trademarks, service marks, etc. in this publication does not imply, even in the absence of a specific statement, that such names are exempt from the relevant protective laws and regulations and therefore free for general use.
The publisher, the authors and the editors are safe to assume that the advice and information in this book are believed to be true and accurate at the date of publication. Neither the publisher nor the authors or the editors give a warranty, expressed or implied, with respect to the material contained herein or for any errors or omissions that may have been made. The publisher remains neutral with regard to jurisdictional claims in published maps and institutional affiliations.

This Springer imprint is published by the registered company Springer Nature Singapore Pte Ltd.
The registered company address is: 152 Beach Road, #21-01/04 Gateway East, Singapore 189721, Singapore

Supervisor's Foreword

The Ph.D. thesis by Xiaoyi Liu addresses some important issues on the nanomechanics of graphene, a subject of importance for a fundamental understanding of graphene deformation and for providing useful guidance for the design of graphene-related materials.

In the past decades, graphene has become increasingly interesting in materials science, physics, chemistry, and mechanics due to its fascinating properties such as extremely high strength, high thermal conductivity, quantum electronic transport, and optical properties. Especially, for materials science and mechanics, understanding the mechanical properties of graphene can also provide some valuable methods and theories for similar low-dimensional materials research.

Although graphene is considered as the strongest material ever known, the two-dimensional (2D) structure limits its applications as structural materials. The strength defined for graphene is generally in its basal plane. Actually, the out-of-plane mechanical behaviors of graphene play a dominant role in its three-dimensional (3D) deformation, which is extremely important for the realistic applications in engineering. On the road to the practical use of graphene, we inevitably need to answer the following questions. How about the out-of-plane mechanical properties of graphene? For 3D graphene composites, how do the out-of-plane mechanical properties of graphene affect the performance of composites? Are there some basic principles of designing graphene and graphene composites?

The nature of the out-of-plane mechanical properties of graphene can open up a new era of multiscale mechanics. For the common 3D materials, the effects of inner interfaces such as grain boundaries are complicated due to the spatial microstructure. Quasi-3D theoretical mechanical models can partially solve this problem. However, most of these models are difficult to be verified by experiments due to the 3D microstructures of materials. The multiscale models to characterize the mechanical properties of graphene can be directly verified by experimental results, which may also be applicable to other nanomaterials with similar structures.

Most of the materials on earth are consisting of discrete atoms. Therefore, the continuum mechanical theories cannot be applicable under any circumstances. Through studying the discontinuous effects on the out-of-plane mechanical properties of graphene, it is possible to provide some supplements to the mechanical theories in this respect, which is especially important for the mechanical behaviors of materials at nanoscale.

In recent years, experimental and theoretical studies show that the mechanical properties of these composites are influenced by graphene, matrix materials, and their interface characteristics. Based on the out-of-plane mechanical behavior of graphene and discontinuous effects, the underlying mechanisms of strengthening effects of graphene in graphene-enhanced composites should be revealed, which is important for the design of graphene composites and other nanomaterials with similar structures.

In order to resolve these issues listed above, the thesis by Xiaoyi Liu presents a systematic study on the fundamental mechanical properties of graphene and graphene composites, and their potential applications in engineering, largely with molecular dynamics simulations and theoretical modeling. It focuses on the out-of-plane mechanical behaviors of graphene, and their effects on the mechanical properties of graphene composites. The results may provide useful guidance for the design of graphene-related materials. This thesis is of interest to students and scientists working in mechanics, condensed matter physics, and materials science.

Hefei, China
April 2019

Hengan Wu

Abstract

Graphene is a two-dimensional material consisting of a monolayer of carbon atoms. Due to its fascinating properties such as extremely high strength, high thermal conductivity, quantum electronic transport, and optical properties, graphene has become increasingly interesting in materials science, physics, chemistry, and mechanics. The physical properties of graphene such as electronic, magnetic, and optical properties can be tailored by its deformation, and thus bear potential applications in, *e.g.*, nanomaterials and nanodevices in nanoelectromechanical systems. Compared with common materials, graphene can be designed at atomistic scale to synthesize desired functionalized graphene and graphene-improved composites. The mechanical behaviors of graphene in these materials play a dominant role in their electromechanical performances. Understanding the mechanical properties of graphene is the foundation of designing graphene and graphene composites.

Given its two-dimensional structure, the mechanical properties of graphene are generally classified as in-plane and out-of-plane ones. Compared with in-plane deformation, the physical properties of graphene due to out-of-plane deformation are more tunable. In addition, graphene generally exists as interfaces in composites, and its out-of-plane deformation plays a dominant role in the mechanical performance of composites. This thesis is aiming at studying some important and unresolved issues in this field: the out-of-plane mechanical properties of graphene and the related effects on the design of graphene composites are studied systematically.

The basic mechanisms of static and dynamic out-of-plane mechanical properties are revealed by studying buckling and transverse waves in graphene monolayer. Due to the discontinuous effects induced by electron density and warped edge, the buckling in graphene monolayer is chirality and size dependent. The propagation of transverse waves in graphene monolayer becomes anisotropic as a result of the chiral difference in bending stiffness when the vibrational frequency is over 3 THz. In addition, discontinuous effects occur when the wavelength is close to the lattice size of graphene, giving rise to anisotropic upper limitations of frequency for transverse waves. With the discontinuous effects and the anisotropy considered, the applicable frequency of transverse waves in graphene monolayer ranges from 1

THz to 3 THz to achieve improved electronic properties in nanoelectromechanical systems.

In graphene, one type of defects is grain boundaries as in 3D materials. Inducing defects into graphene is considered as the foundation of the atomistic design. As a common method to induce defects, the bombardment of a suspended monolayer graphene sheet via different energetic atoms is investigated. As the electronegativity of incident atom increases, the applicable incident energy range for one-step substitution widens, while that for generating single-vacancy narrows. The mechanical behaviors of graphene nanoribbon under torsion loading, which is a representative case including both static and dynamic out-of-plane deformation, are analyzed to reveal the discontinuous effects induced by defects. In general, defects decrease the characteristic size of graphene, leading to an increase in the discontinuous effects. Introducing defects into graphene can be beneficial to mechanical and electronic properties simultaneously.

Graphene/matrix interface plays a dominant role in the mechanical properties of graphene composites. Based on the understanding of out-of-plane mechanical behaviors, the mechanical properties of graphene nanolayered composites are investigated under out-of-plane shock and in-plane shear loading. The out-of-plane deformation of graphene interfaces in nanolayered composites deteriorates the in-plane strength of composites, but improves the out-of-plane strength. The mechanisms of blocking and healing dislocations of graphene interfaces in nanolayered composites are also revealed.

In summary, the out-of-plane mechanical properties of graphene and its influence on the mechanical properties of graphene composites are investigated. I expect the results from this thesis to provide a valuable guideline for design and application of graphene and graphene composites in the field of nanomechanics.

List of publications

Journal publications (in chronological order)

1. X. Y. Liu, F. C. Wang* and H. A. Wu*. 'Anisotropic propagation and upper frequency limitation of terahertz waves in graphene.' *Applied Physics Letters*, 2013. 103(7): 071904.

2. X. Y. Liu, F. C. Wang, H. S. Park and H. A. Wu*. 'Defecting controllability of bombarding graphene with different energetic atoms via reactive force field model.' *Journal of Applied Physics*, 2013. 114(5): 054313.

3. X. Y. Liu, X. Z. Xu*. 'Mesoscopic numerical computation model of air-diffusion electrode of metal/air batteries.' *Applied Mathematics and Mechanics (English Edition)*, 2013. 34(5): 571–576.

4. X. Y. Liu, F. C. Wang, H. A. Wu* and W. Q. Wang*. 'Strengthening metal nanolaminates under shock compression through dual effect of strong and weak graphene interface.' *Applied Physics Letters*, 2014. 104(23): 231901.

5. X. Y. Liu, F. C. Wang and H. A. Wu*. 'Anisotropic growth of buckling-driven wrinkles in graphene monolayer.' *Nanotechnology*, 2015. 26(6): 065701. (*Selected as feature article and cover page*).

6. Q. L. Wei, S. S. Tan, X. Y. Liu, *et al*. 'Novel polygonal vanadium oxide nanoscrolls as stable cathode for lithium storage.' *Advanced Functional Materials*, 2015. 25(12): 1773–1779.

7. S. S. Lin*, S. J. Zhang, X. Q. Li, W. L. Xu, X. D. Pi, X. Y. Liu, *et al*. 'Quasi-Two-Dimensional SiC and SiC2: Interaction of Silicon and Carbon at Atomic Thin Lattice Plane.' *Journal of Physical Chemistry C*, 2015. 119(34): 19772–19779.

8. X. Y. Liu, F. C. Wang* and H. A. Wu. 'Anomalous twisting strength of tilt grain boundaries in armchair graphene nanoribbons.' *Physical Chemistry Chemical Physics*, 2015. 17(47): 31911–31916.

9. H. A. Wu* and X. Y. Liu. 'Tuning electromechanics of dynamic ripple pattern in graphene monolayer.' *Carbon*, 2016. 98: 510–518.

10. S. J. Zhang, S. S. Lin*, X. Q. Li, X. Y. Liu, *et al*. 'Opening the band gap of graphene through silicon doping for the improved performance of graphene/GaAs heterojunction solar cells.' *Nanoscale*, 2016, 8(1): 226–232.

11. X. Y. Liu, F. C. Wang*, W. Q. Wang* and H. A. Wu. 'Interfacial strengthening and self-healing effect in graphene-copper nanolayered composites under shear deformation.' *Carbon*, 2016. 107: 680–688.

Review

X. Y. Liu, F. C. Wang, H. A. Wu*. 'Research progress in nanomechanics of graphene and its composites.' *Chinese Journal of Solid Mechanics*, 2016, 37(5): 398–420. (Invited review)

Conferences

1. X. Y. Liu, oral report: 'Defecting controllability of bombarding graphene with single energetic atoms.' *The Chinese Society of Theoretical Applied Mechanics Conference*, Xi'an, Aug 2013: MS5905.

2. X. Y. Liu, oral report: 'Atomic Analysis of Dwell-penetration Transition Velocity for Oblique Targets.' *Asia–Pacific Congress for Computational Mechanics*, Singapore, 2013: 1158.

3. X. Y. Liu, invited and keynote report: 'Chirality-dependent buckling-driven wrinkles in graphene monolayer.' 6th *International Conference on Computational Methods*, Auckland, 2015: 683.

4. X. Y. Liu, oral report: 'Anisotropic buckling properties of graphene monolayer.' *The Chinese Society of Theoretical Applied Mechanics Conference*, Shanghai, Aug 2015: MS8107.

Acknowledgements

This thesis was finished under the guidance of my advisor, Prof. HengAn Wu, at the Computational Mechanics (CME) lab, Department of Modern Mechanics, University of Science and Technology of China (USTC). During the 5 years at the CME lab, Prof. Wu helped me tremendously in both study and daily life. Joining his group is a major turning point in my life. His profound knowledge, broad perspective, rigorous scholarship, and pursuit of excellence are extremely impressive, and have impacted my academic career. I felt very fortunate to have such an excellent supervisor to whom I am indebted greatly.

When I was a freshman in the CME lab, Prof. Xiu-Xi Wang also helped me in various ways, and offered valuable advice later on regarding my research including meticulous proofreading of this thesis. Dr. Fengchao Wang taught me molecular dynamics and provided considerable help in my research. I am thankful for their help and care.

I also had the pleasure to learn dislocation analysis from Profs. Linghui He and Yong Ni. The initial idea of studying graphene composites was inspired by Prof. Wengqiang Wang from Institute of Fluid Physics, China Academy of Engineering Physics. Dr. Huimin Li in Supercomputing Center at USTC helped me with supercomputing matters. I am extremely grateful for all their help.

I would like to thank my fellow students at USTC. Haoran Liu offered his ample help in my study and daily life. I discussed shock loading with Bo Li, Yang Cai, and Fengpeng Zhao, and learned a lot from them. It was fruitful and pleasant to discuss with Yongkuan Shen, Xiaolong Wang, Tianwu Zhao, Jie Chen, Yingqi Li, Liya Wang, Chuang Liu, and Yinbo Zhu in the CME lab. Jun Xia and Yongchao Wang also helped with proofreading this thesis. The wonderful studying and working atmosphere in the CME lab allowed me to work efficiently and finish my Ph.D. study.

The English version of this thesis was written at the Peac Institute of Multiscale Sciences (PIMS), Chengdu, where I have been a junior faculty member since 2016. PIMS is a research and education institute with focus on understanding materials properties at multiple temporal and spatial scales, in particular at submicron and subnanosecond scales. The subject areas span over materials science, mechanics, condensed matter and plasma physics, chemistry, optics including THz and X-ray sciences. PIMS Director, Prof. Sheng-Nian Luo, gave me freedom to compose this English version and to continue my research on nanomechanics, and helped with English writing. I am grateful for the generous support from PIMS.

Finally, I would like to thank my parents for their continued support and love, which encourage me to move forward and finish my education.

Chengdu, China
April 2019

Xiaoyi Liu

Contents

1 Introduction 1
   1.1 Graphene and Graphene Composites 1
   1.2 Brief Review on Nanomechanics of Graphene
       and Graphene Composites 2
       1.2.1 In-Plane Mechanical Properties of Graphene 3
       1.2.2 Out-of-Plane Mechanical Properties of Graphene 6
       1.2.3 Influence of Defects on Mechanical Properties
            of Graphene 9
       1.2.4 Mechanical Properties of Graphene Composites 10
   1.3 Some Important Issues in this Field 13
   1.4 Thesis Overview 15
   References 15

2 Multiscale Methods to Investigate Mechanical Properties
   of Graphene 19
   2.1 Introduction 19
   2.2 MD and DFT Simulation Methods 21
       2.2.1 MD Simulation 21
       2.2.2 DFT Simulation 24
   2.3 Conversion of Physical Quantities from MD Simulations
       to Continuum Models 25
   2.4 Chapter Summary 27
   References 27

3 Buckling of Graphene Monolayer Under In-Plane Compression ... 29
   3.1 Introduction 29
   3.2 Simulation Details 30
       3.2.1 MD Simulation 30
       3.2.2 DFT Simulation 31

	3.3	The Buckling of Graphene Monolayer Under Hydrostatic Stress	31
	3.4	Discontinuous Effects in the Buckling of Graphene Monolayer	32
	3.5	The Buckling of Graphene Without Discontinuous Effects	35
	3.6	Chapter Summary	37
	References		37
4	**Dynamic Ripples in Graphene Monolayer**		**39**
	4.1	Introduction	39
	4.2	Transverse Waves in Graphene Monolayer	41
		4.2.1 Propagation of Transverse Wave in a Square Graphene Sheet	41
		4.2.2 The Effects of Chirality and Vibrational Frequency on Transverse Waves Propagation	42
		4.2.3 Discontinuous Effects on the Permissible Frequency	43
	4.3	Controllable Dynamic Ripples in Graphene Monolayer	45
		4.3.1 Simulation Details	45
		4.3.2 Patterns and Quality of Dynamic Ripples Generated by the Interference of Transverse Waves	46
		4.3.3 The Motion and Electronic Properties of Dynamic Ripples	49
	4.4	Chapter Summary	52
	References		53
5	**Defect-Induced Discontinuous Effects in Graphene Nanoribbon Under Torsion Loading**		**55**
	5.1	Introduction	55
	5.2	The Defect Controllability of Graphene Under Atomistic Bombardment	56
		5.2.1 MD Simulation Details	56
		5.2.2 The Effect of Impact Site	57
		5.2.3 The Effects of Physical Properties and Kinetic Energy of Incident Atom	58
	5.3	Graphene Nanoribbon with Grain Boundary Under Torsion Loading	61
		5.3.1 MD Simulation Details	61
		5.3.2 The Instability of GNRs Under Torsion Loading	63
		5.3.3 The Electron Transport Properties of tAGNRs	66
	5.4	Chapter Summary	68
	References		68
6	**Mechanical Behaviors of Graphene Nanolayered Composites**		**71**
	6.1	Introduction	71
	6.2	The Strong/Weak Duality of Graphene Interfaces Under Out-of-Plane Shock Loading	73

		6.2.1	Models and Methods	73
		6.2.2	The Strengthening Effects of Graphene Under Bullet Impact	74
		6.2.3	The Effects of Graphene Interface on Shock Waves	75
	6.3	\multicolumn{2}{l}{The Mechanical Properties of GCuNL Composites Under In-Plane Shear Loading}	78
		6.3.1	Models and Methods	78
		6.3.2	Shear Responses of GCuNL Composites	80
		6.3.3	The Yield and Failure of GCuNL Composites	81
		6.3.4	Self-Healing Effects in GCuNL Composites	83
	6.4	\multicolumn{2}{l}{Chapter Summary}	86
	\multicolumn{3}{l}{References}	86
7	\multicolumn{3}{l}{**Summary and Future Work**}	89
	7.1	\multicolumn{2}{l}{Summary}	89
	7.2	\multicolumn{2}{l}{Novelty}	90
	7.3	\multicolumn{2}{l}{Future Work}	91

Appendix A: C++ Codes to Model Grain Boundaries in Graphene Monolayer ... 93

Appendix B: C++ Codes for Binning Analysis ... 97

Chapter 1
Introduction

1.1 Graphene and Graphene Composites

Graphene was first separated from graphite by Geim et al. using an adhesive tape in 2004 [1, 2]. It is a two-dimensional (2D) material with a single layer of carbon atoms arranged in a hexagonal lattice, and opens up a new era of 2D materials. Though graphene is a 2D material with a monoatomic thickness, its strength is extremely high. Graphene monolayer consists of hexagonal lattices in a plane (Fig. 1.1), and the carbon–carbon bonds with a length of 1.42 Å form via sp^2 orbital hybridization. It becomes a hotspot in physics, materials science, chemistry and mechanics due to its fascinating properties such as electromagnetic [3], thermal [4], optical [5] and mechanical [6] properties.

Recent studies demonstrate a great potential of graphene in such applications as nanoelectronic devices [7], composites [8], nanoelectromechanical systems (NEMS) [9] and energy storage [10]. For example, the electromagnetic properties of graphene can be tuned by deformation, and this "mechanoelectromagnetism" coupling effect has important potential applications in NEMS [11, 12]. In addition to the electromagnetic properties, the mechanical properties of graphene have become increasingly interesting in recent years. The nanoindentation experiments performed by Lee et al. show that graphene has a Young's modulus of about 1 TPa, and a tensile strength of about 130 GPa [6], and is the strongest material ever known. Due to its unique atomistic structure, graphene can be designed at nanoscale to tune its mechanical properties [13]. Understanding its mechanical properties is the foundation of designing graphene materials with high performances [14].

On the other hand, the 2D structure of graphene limits its applications in engineering as a structural material. Fortunately, due to its high strength and toughness, graphene is considered as a promising strength enhancer in composites [15]. Recent research results show that doping graphene into composites can improve chemical [16], electromagnetic [17], thermal [18] and mechanical [19] properties of materials. In addition, the mechanical behaviors of graphene play a dominant role in the mechanical properties of three-dimensional (3D) graphene composites [20].

© Springer Nature Singapore Pte Ltd. 2019
X. Liu, *Nanomechanics of Graphene and Design of Graphene Composites*,
Springer Theses, https://doi.org/10.1007/978-981-13-8703-6_1

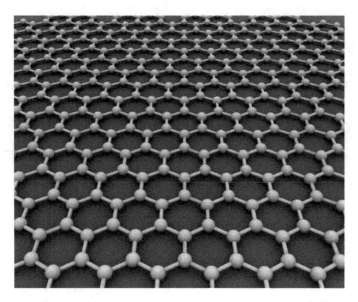

Fig. 1.1 Atomistic configuration of graphene

Therefore, understanding mechanical properties of graphene is a prerequisite for designing and fabricating graphene-related materials. This book is intended to explore mechanical properties of graphene and its mechanical behaviors in composites, summarize some primary principles of designing graphene and its composites, and thus provide certain theoretical guidelines for materials design at nanoscale, as well as engineering applications of graphene and its composites. In this chapter, the studies on mechanical properties of graphene and its composites in recent years are reviewed, and some important issues are raised.

1.2 Brief Review on Nanomechanics of Graphene and Graphene Composites

Compression, tension, torsion and bending are four fundamental ways of loading for general 3D materials (Fig. 1.2a). Due to its 2D structure, the mechanical properties of graphene are classified as in-plane and out-of-plane properties (Fig. 1.2b). In general, the in-plane properties are concerned with deformation or loading in the basal plane of graphene, while out-of-plane properties, with deformation or loading out of the basal plane. Previous studies on the in-plane mechanical properties of graphene investigated Young's modulus, Poisson's ratio, fracture, and friction under in-plane loading. Compared with in-plane mechanical behaviors, the out-of-plane mechanical behaviors are richer and more complicated, and of more interest in recent years (e.g., the fracture under out-of-plane ballistic impact). Moreover, understanding the

1.2 Brief Review on Nanomechanics of Graphene and Graphene Composites

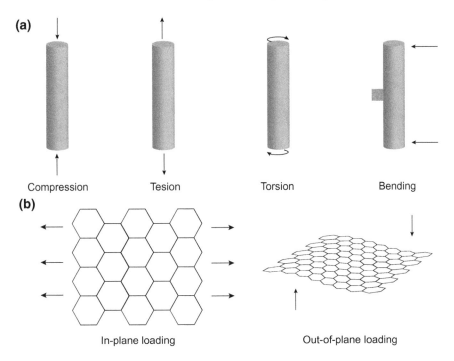

Fig. 1.2 a Common loading modes for 3D materials: compression, tension, torsion and bending. **b** In-plane and out-of-plane loading of graphene

mechanical behaviors of graphene in composites and related mechanisms is of great importance for improving the mechanical properties of graphene composites, and is also a basis for related materials design.

1.2.1 In-Plane Mechanical Properties of Graphene

Given graphene's monoatomic-layer structure, it is challenging to stretch graphene in-plane directly. One of the common and simple methods to obtain the in-plane mechanical behaviors of graphene is nanoindentation (Fig. 1.3): graphene is fixed on a substrate with holes, and then the out-of-plane loading is applied via nanoindentation. The in-plane tensile mechanical properties of graphene are indirectly characterized via measuring the relation between the displacement of the indenter and the force. In 2008, the in-plane tensile properties of graphene were measured through the nanoindentation experiments by Lee et al. Their results show that, the Young's modulus of graphene is about 1.0 ± 0.1 TPa, and its strength is about 130 ± 10 GPa. Using an atomic force microscope (AFM), Frank et al. [21] found that the spring constants of graphene ranges from 1 to 5 N/m, the equivalent Young's modulus of

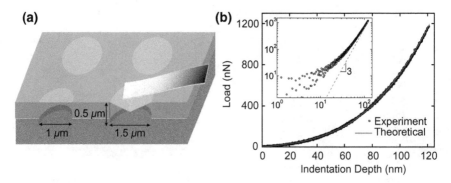

Fig. 1.3 a Schematic of nanoindent experiments on graphene. **b** Indentation depth versus load curves. The origin data is exported from Ref. [6]

doubly clamped beams under tension is 0.5 TPa, and the in-plane Young's modulus of graphene is about 1.0 TPa. The Young's modulus and strength of graphene are much higher than those of other known materials. This is a main reason why graphene is of significant interest in the field of mechanics.

In addition to experimental efforts, theoretical modeling and numerical simulations have also been conducted extensively to study the mechanical properties of graphene in recent years. One of the most common continuum mechanics models for graphene is the thin shell model [22]. However, the thickness of graphene thickness is not well defined, which may be different in different continuum mechanics models [23]. For most continuum mechanics models, the thickness of graphene is defined as $\sim$0.34 nm. In 2007, Li et al. [24] used first-principles calculations to study the ideal strength and phonon instability of graphene under tension. Their results show that the Young's modulus and the Poisson's ratio of graphene are 1,050 GPa and 0.186, respectively. The phonon instability point of graphene is chirality-dependent: it occurs at ($\varepsilon_{xx} = 0.194$, $\sigma_{xx} = 110$ GPa, $\varepsilon_{yy} = -0.016$) along the armchair direction, but at ($\varepsilon_{yy} = 0.266$, $\sigma_{xx} = 121$ GPa, $\varepsilon_{xx} = -0.027$) along the zigzag direction.

As early as 2000, Van Lier et al. [25] used first-principles calculation to predict the elastic mechanical properties of monolayer graphite (graphene): they assumed that the thickness of monolayer graphite to be 0.34 nm, and its tensile Young's modulus was calculated to be 1.11 TPa. The results of Konstantinova et al. [26] from first-principles calculations show that, the tensile Young's modulus of graphene monolayer is 1.25 TPa and 1.23 TPa for the LDA and GGA methods, respectively. Regardless of the thickness, there are also some methods of characterizing the tensile mechanical properties of graphene by directly using strain energy corresponding to the second derivative ($d^2 E/d\varepsilon^2$): the theoretical value of $d^2 E/d\varepsilon^2$ obtained by Sánchez-Portal et al. [27] is about 60 eV, and that by Kuding et al. [28] is about 57.3 eV.

In 2009, Zhao et al. [29] used molecular dynamics (MD) simulations to study the fracture of graphene under tensile loading. Their results (Fig. 1.4) show that there is no chiral difference for the in-plane tensile mechanical properties of graphene at small strains, while at large strains they become chirality-dependent: the tensile

1.2 Brief Review on Nanomechanics of Graphene and Graphene Composites

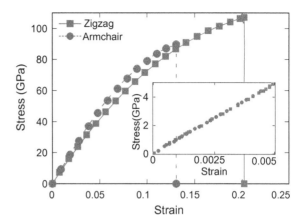

Fig. 1.4 Chirality-dependent tensile strength of single-crystal graphene. The origin data is exported from Ref. [29]

strength along the zigzag direction is higher than that along the armchair direction. Aluru et al. further studied the effects of temperature and strain rate on graphene fracture [30]. Their results show that the fracture strength of graphene decreases with the increase of temperature, and is not affected by strain rate.

Although previous experiments and simulations show that graphene has a very high Young's modulus and strength, Zhang et al. proposed a new perspective in 2014 on the strength and toughness of graphene [31]: the strength and toughness of graphene is generally considered very strong, but their experimental results show that graphene is actually a brittle material with a low strength and toughness: the fracture toughness of graphene is only $4.0 \pm 0.6\,\mathrm{MPa\,m^{1/2}}$, and the equivalent critical strain energy release rate is only $15.9\,\mathrm{J\,m^{-2}}$. They also studied dynamic mechanical behavior of graphene under ballistic impact, and found that although the fracture strength of graphene is not as high as theoretical predictions, the energy delocalization effect due to high speed of sound leads to a specific penetration energy for multilayer graphene 10 times higher than steel [32].

Continuum mechanics models of graphene (e.g., thin membrane and shell models) are not applicable in all cases. In 2012, Gao et al. studied tensile fracture of polycrystalline graphene with a small hole [33]. According to the theoretical solution of stress concentration in a plate with a small hole, the stresses at the upper and lower edges of the small hole are 3 times those at the far field. Therefore, initial cracks are generated at the upper or lower edges of the small hole. However, Gao et al. showed that initial cracks in polycrystalline graphene are not generated at the upper or lower edges of the hole, but are generated in the far field (Fig. 1.5). This indicates that, although in most cases, continuum mechanics models (e.g., membrane and shell) can well describe the in-plane mechanical behaviors of graphene, they are not applicable in all cases. At nanoscale, the discontinuous effects should be considered, which will be further discussed as an important issue in following chapters.

In addition, in-plane slip (friction) of graphene has also become an interesting topic in recent years, since multilayer graphene exhibits superlubricity characteristics

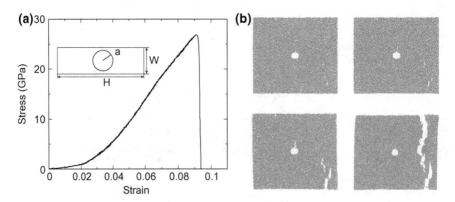

Fig. 1.5 Tensile fracture of polycrystalline graphene sheet with a small hole. **a** Stress–strain curve. Inset: geometry of a polycrystalline graphene sheet. The origin data is exported from Ref. [33]. **b** Snapshots of graphene fracture obtained from MD simulations in this thesis

[34, 35]. Zheng et al. conducted in-depth research on the superlubricity of graphite and multilayer graphene. They studied the self-recovering motion of graphite sheets and multilayer graphene, achieved superlubricity at micrometer scale under non-vacuum conditions [36–38], and found that the cleavage energy of graphite was not influenced by temperature, twist angle, scale and vacuum [39].

1.2.2 Out-of-Plane Mechanical Properties of Graphene

Common out-of-plane deformations of graphene are wrinkling and rippling. Ripples in graphene can be static [41] or dynamic [42]. Zan et al. observed a large number of wrinkles in graphene monolayer at room temperature with a scanning tunneling microscope (Fig. 1.6) [40]. 2D crystals are generally considered unstable according to thermodynamics; however, these wrinkles are the dominant factor in maintaining the stability of graphene [43, 44]. Compared with in-plane deformation, the physical properties of graphene due to out-of-plane deformation are more tuneable [45, 46]. For example, although the electron mobility and corresponding resistivity of graphene is 15,000 cm^2/V s and 10^{-6} Ω cm respectively, its band gap is 0 [2]. Therefore, graphene cannot be directly used as semiconductors in electronic devices (the current common material is silicon) [47]. Out-of-plane deformation can open the band gap of graphene while retaining its electron mobility [48]. In addition, graphene has low bending stiffness and high ductility, which are desirable for the development of wearable electronic devices which may undergo large deformation [49].

Although electronic properties of graphene can be modified by out-of-plane deformation, only controllable out-of-plane deformation is useful in a nanoelectromechanical system. In 2009, Lau et al. showed that, controlled rippling can be achieved by

1.2 Brief Review on Nanomechanics of Graphene and Graphene Composites

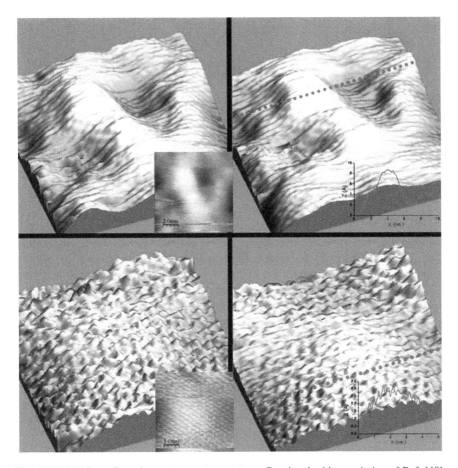

Fig. 1.6 Wrinkling of graphene at room temperature. Reprinted with permission of Ref. [40], copyright 2012, Royal Society of Chemistry

sandwiching a graphene sheet on two substrates with a temperature difference to generate thermal stresses (Fig. 1.7) [12]. San et al. found that ripples can be induced by electron excitation [50]. Singh et al. demonstrated that ripples in graphene can be controlled by temperature, strain, and layer numbers, and the scaling relation for average angular deviations depends on graphene sheet size and averaging radius [51]. Osvath et al. reported that ripples in graphene on a metal substrates can be controlled by electric field [52]. Through simulating graphene buckling under shear loading, Wang et al. found that periodic wrinkles can be controlled by shear strain and the patterns of wrinkles are influenced by graphene size [53].

Compared with free-standing graphene, out-of-plane deformation of graphene on substrates is easier to control. In 2008, Bunch et al. placed graphene on a SiO_2 substrate with nanovoids to form a gas chamber, and then applied pressure to the gas chamber to obtain the elastic constants of graphene from the relation between

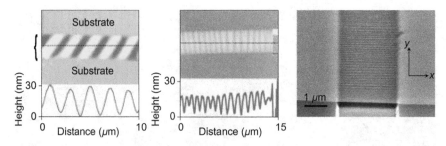

Fig. 1.7 Controlled rippling in graphene monolayer induced by thermal stress. Reprinted with permission of Ref. [12], copyright 2009, Nature Publishing Group

out-of-plane displacement and applied pressure [9]. It was found that an ordinary gas cannot penetrate graphene, and stable ripples in graphene can be controlled through such a gas chamber. Out-of-plane deformation can also be generated by placing graphene on metallic substrates. Parga et al. grew epitaxial graphene monolayers on Ru(0001) surfaces, and the weakly coupled graphene monolayer was periodically rippled and it exhibited inhomogeneous charge distribution [48]. Paronyan et al. grew graphene on a copper foil by chemical vapor deposition (CVD) from methane, and obtained stable ripples upon thermal quenching from elevated temperatures [54]. Similar results were obtained by Gao et al. through growing graphene on Cu(111) [55]. Locatelli et al. investigated graphene on square Ir(100) substrates, and found that distinct physisorbed and chemisorbed graphene phases coexisted on the surface below 500 °C, respectively characterized by flat and buckled morphology [56]. Rippling in graphene bilayers was reported by Mao et al. in 2011, who considered the graphene bilayers as inter-substrates [57].

Similar to stable ripples, dynamic ripples in graphene have also attracted considerable attention in recent years. Dynamic ripples in graphene, also known as transverse waves, are generally obtained by nanoindentation through periodic out-of-plane loading [58, 59]. Park et al. studied transverse waves in graphene monolayer by MD simulations and continuum mechanics theories, and obtained the effective thickness of graphene [60]. In 2008, the concept of graphene-based electromechanical nanodevices was proposed by Garcia-Sanchez et al. [61]. Due to graphene's low density and high strength, the applicable frequency of transverse wave is on the order of terahertz, which is considered as a promising nanoresonators [62]. However, compared with stable ripples, systematic studies are few on the fundamental mechanisms of dynamic ripples in graphene, which will be elaborated in the following chapters.

1.2.3 Influence of Defects on Mechanical Properties of Graphene

Inducing defects into graphene is a basic way of its atomistic design, which can also improve the electromagnetic properties. Defects can induce magnetic fields and open band gaps in graphene [63, 64]. Doping "metallic" atoms in graphene defects is a typical approach in designing graphene materials on atomistic scale, which can improve the spintronics [65] and electron transport properties of graphene [66, 67]. Inducing defects and their effects on mechanical properties of graphene are two issues that are of great concern in the field of mechanics.

Bombarding nanomaterials with atomic, ion or electron beams is a common method for nano-cutting [68], and has recently been applied to introducing defects into graphene [69–71]. In addition, studying the mechanical behavior of graphene subjected to particle radiation is of great significance for spacecraft protection in outer space [72]. In 2012, Wang et al. studied atom bombardment of graphene, and proposed a two-step process for doping graphene at atomistic scale [73]: create vacancies by high energy atom/ion bombardment, and then fill these vacancies with low energy desired dopants. However, bombardment is a random process, and the probability of defect formation is affected by the energy of an incident atom. Krasheninnikov et al. studied the probability distribution of defects induced by bombardment by nitrogen and helium atoms at different incident energies [74, 75].

The effects of defects on mechanical properties of graphene have also attracted considerable attention in recent years. In 2010, Rhonda et al. showed that randomly distributed defects in graphene do not affect the expansion direction of cracks, while periodically arranged defects highly affect crack propagation direction [76]. Long et al. studied the evolution of defects in graphene under in-plane shock loading [77]. Actually, grain boundaries in graphene are generated by defects arranged in specified paths, which are ubiquitous in realistic polycrystalline graphene. In 2011, Kim et al. realized controllable growth of grain boundaries in graphene by using electron diffraction in a scanning transmission electron microscope (STEM) [78]. In 2010, Ruoff et al. studied the effects of grain boundaries on tensile fracture of graphene by MD simulations, and their results showed that the strength is not affected by defects (Fig. 1.8) [79]. Wei et al. claimed that the strengthening or weakening effects of grain boundary in graphene depends on the arrangement and density of defects [80], and the carbon–carbon bond energy plays a dominant role in the fracture surface energy, which leads to the preferred propagating direction of cracks along the zigzag direction in graphene monolayer [81]. This is similar to the "stress concentration in graphene with small hole" problem mentioned above [33], which cannot be explained by classical theories of continuum mechanics. In addition, as temperature increases, the transition probability of defects increases, which will degrade the strength of graphene [82].

Out-of-plane mechanical properties of graphene can also be affected by defects. In 2012, Neek et al. studied the effects of grain boundary on graphene buckling, and their results showed that buckling strain peaks at grain boundary angle $\theta = 21.8°$,

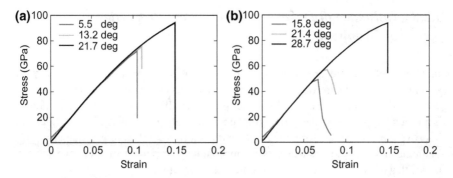

Fig. 1.8 Stress–strain curves of **a** zigzag-oriented and **b** armchair-oriented graphene sheets pulled perpendicular to grain boundaries. The origin data is exported from Ref. [79]

and is the lowest at $\theta = 32.2°$ [83]. Recently, Zhang et al. point out that, the periodic arranged ripples in graphene can be obtained by tailoring defects [84]. Although it is still challenging to control defects, these studies provide a valuable insight for atomistic design of graphene.

There has not been a universal continuum mechanical theory to characterize the effects of defects on mechanical properties of graphene. For different problems, different methods or mechanical models are proposed to reveal the underlying mechanism accordingly. Actually, there exist both continuous and discontinuous effects in graphene. Introducing defects into graphene increases the discontinuous effects, which will be discussed in following chapters.

1.2.4 Mechanical Properties of Graphene Composites

The 2D structure limits the applications of graphene as structural materials. In order to break through this limitation, graphene composites are considered promising in engineering applications [8]. Graphene composites can be classified into two types according to matrix materials: for the first type, graphene is used as matrix material, which is aimed at designing graphene at nanoscale to improve its electromagnetic properties (e.g., the magnetism [63] and electron transport properties [66, 67]); for the second type, graphene is used as an additive to improve physical properties of matrix materials. The former is called functionalized graphene, and the latter is called graphene-improved composites. In general, "graphene composites" refers to the latter unless stated otherwise.

The synthesis of functionalized graphene is generally achieved by doping other atoms into vacancies and other defects in graphene. One representative type of functionalized graphene is graphene oxide [85–89]. Yu et al. conducted a series of studies on the synthesis of graphene oxide. In 2012, they reported a one-step fabrication of macroscopic multifunctional graphene-based hydrogels with robust interconnected

1.2 Brief Review on Nanomechanics of Graphene and Graphene Composites

networks via reducing graphene oxide sheets by ferrous ions and in situ simultaneous depositing nanoparticles on graphene sheets [90]. Moreover, they synthesized a graphene oxide fiber with a tensile strength of 125 MPa [91]. In addition, they doped silver atoms in a graphene oxide sheet to form an antibacterial composite, which can be used as a biomedical material [92]. In 2013, they reported flexible graphene-polyaniline composite paper for high-performance supercapacitor without losing the electrical properties of graphene, which is of great promise for the development of low-cost electrode materials in energy storage devices [93].

Since graphene is used as a matrix material, the functionalized graphene is generally 2D materials. On the other hand, the graphene-improved composites are generally 3D materials, which is also referred to as graphene composites as mentioned above. In the field of mechanics, graphene composites are currently more concerned, and the key idea is to introduce graphene as an enhancer in composites [15]. Random doping of graphene into a metallic matrix is a common and simple approach to fabricating graphene composites. Rafiee et al. conducted a series of research on this issue. In 2010, graphene nanoribbon composites were synthesized by them [94], and the Young's modulus and ultimate tensile strength of the composites are increased by 30% and 22% with a content of 0.3% graphene, respectively. In 2011, graphene–ceramic composites were synthesized, and their fracture toughness was improved by ~235% [95]. Tian et al. fabricated cellulose/graphene composite fibers with a 25% increase in the Young's modulus and a 50% increase in fracture strength with only 0.2 wt% graphene [96]. However, the strengthening mechanisms of graphene in these composites at nanoscale are still unclear, and not all random doping can increase the strength of the composites. For example, Bartolucci et al. showed that the strength and toughness of graphene–aluminum nanocomposites are less than that of pure aluminum [97]. A mainstream view of strengthening mechanism of graphene in composites is based on the interaction between carbon and matrix atoms [98, 99]: chemical bonding leads to a significant deterioration in strength, while van der Waals forces can significantly improve the strength.

Although a number of experimental results are consistent with this mechanism, its guiding significance for designing graphene composites is still limited. Actually, the proportion of graphene can be precisely controlled, while it is still challenging to synthesize composites with desired microstructures. Even if the proportion of doped graphene is identical, graphene composites with different microstructures will probably perform differently [100]. In 2013, Geim et al. proposed the concept of graphene heterojunction composites, i.e., composites are formed by adhering different layered nanomaterials together by interlayer van der Waals forces [101]. This concept was realized by Kim et al. [102]: graphene–metal nanolayered (GMNL) composites were fabricated by combining copper/nickel nanolayers with single-layer graphene (Fig. 1.9). Compared with the metallic matrix materials, the strength of graphene-metal nanolayered composites is increased by more than 20 times.

Goli et al. showed that the thermal conductivity of GMNL composites is greatly improved, which is of great significance for the inhibition of adiabatic shear bands [103]. In 2016, Zhang et al. adopted graphene nanoplatelets and reduced graphene oxide to fabricate copper matrix composites through a modified molecular-level

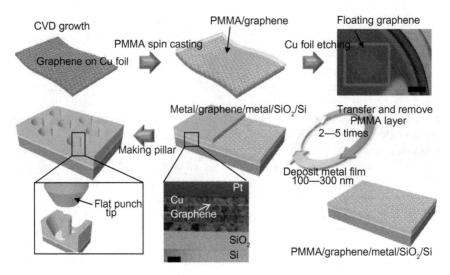

Fig. 1.9 Synthesis scheme of metalgraphene multilayer systems. Reprinted with permission of Ref. [102], copyright 2013, Nature Publishing Group

mixing process, and demonstrated considerable improvement in the mechanical properties of the composite [104]. Hong et al. fabricated cellular micropatterns on a monolayered graphene film with ordered interfaces [105]. Although synthesis of 3D graphene framework composites becomes more and more mature, the matrix material is generally soft polymers or fibers with a relatively low modulus and strength [106].

There are at least two strengthening effects in graphene composites: graphene is in-plane loaded to take advantage of its high strength [102]; graphene interfaces induce self-healing [107] and block dislocations [108]. Similar nanolayered materials may be strengthened via similar mechanisms. Liu et al. showed that fracture toughness of a graphene–ceramic composite is increased by 27.2% under in-plane tension [109]. In 2014, Roman et al. found that the strength of graphene–copper nanolayered composites is also improved considerably [110]. Fang et al. reported that polystyrene nanocomposites with 0.9 wt% graphene nanosheets resulted around 70 and 57% increases in tensile strength and Young's modulus [111]. However, experiments suggest that the mechanical properties of graphene composites deteriorate rapidly when the number of graphene layers is over 5 due to in-plane instability, and composites with graphene monolayers are considered to be promising and applicable in engineering [112, 113].

The strengthening mechanisms of graphene interfaces in composite materials remain unclear. The mechanical properties of graphene composites depend not only on graphene, but also on the interaction between graphene and matrix atoms [114], which is also an important issue discussed in following chapters.

1.3 Some Important Issues in this Field

The brief review on the mechanical properties of graphene and its composites in Sect. 1.2 suggests that in-plane mechanical properties of graphene have been studied extensively and well understood. Compared with in-plane deformation, out-of-plane deformation of graphene is more complex, and there are more unsolved problems. Since most structural materials in engineering applications are three-dimensional, the out-of-plane mechanical properties of graphene play a key role in designing graphene and graphene composites for engineering applications. In this section, some important issues will be addressed.

Out-of-plane deformation of graphene can be induced by both in-plane and out-of-plane loading. In addition, out-of-plane deformation can be static, quasi-static and dynamic. There are four types of out-of-plane deformation for graphene (Table 1.1 lists some representative mechanical behaviors). In particular, the mechanical behaviors under static out-of-plane loading (e.g., nanoindentation) can be considered as a special means to study in-plane mechanical properties of graphene.

According to this classification and the review in Sect. 1.2, the three issues summarized below are important for the design of graphene and its composites.

- The in-plane mechanical properties of graphene have been well studied experimentally and theoretically, but systematic studies are still lacking for graphene's out-of-plane mechanical properties. In particular, buckling and transverse waves in graphene are still unclear, despite their importance for nanoelectronic devices.
- Recent theoretical and experimental studies have shown that discontinuous effects highly affect in-plane mechanical properties of graphene in some specific cases. However, discontinuous effects on out-of-plane mechanical behaviors of graphene have rarely been investigated.
- Out-of-plane deformation of graphene plays a dominant role in the mechanical properties of graphene composites, while the related mechanisms are still poorly understood.

Understanding the fundamental mechanisms underlying out-of-plane deformation of graphene is important for resolving all the three issues listed above. In order to achieve this goal, this thesis addresses three main aspects (Fig. 1.10):

- **Out-of-plane mechanical properties of graphene**.
  There are a variety of types of out-of-plane deformation, and cannot be all discussed in this thesis. However, the basic flexural properties of graphene can be obtained by studying the most representative types including static and dynamic ripples, buckling and transverse waves.

Table 1.1 Four types of out-of-plane mechanical behaviors for graphene

	Static/quasi-static	Dynamic
In-plane loading	Buckling	Dynamic ripples/wrinkles
Out-of-plane loading	Nanoindentation	Vibration and shock

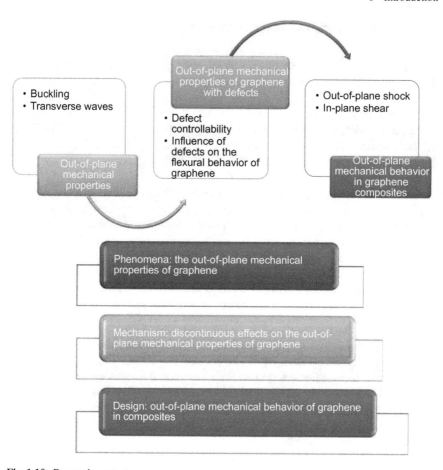

Fig. 1.10 Research contents

- **Defects effects on the mechanical properties of graphene**.
  In general, there are intrinsic discontinuous effects due to the discrete atomistic system of graphene. Defects in graphene may intensify these effects. Compared with out-of-plane mechanical behaviors of pristine graphene, the effects of defects can be revealed, and are important for the design of graphene and functionalized graphene.
- **Mechanical behaviors of graphene in composites**.
  In solid mechanics, graphene-enhanced composites are the most concerned. Experimental and theoretical studies show that the mechanical properties of these composites are influenced by graphene, matrix materials and their interface characteristics. Based on the out-of-plane mechanical behavior of graphene and discontinuous effects, the interfacial behaviors of graphene in graphene-enhanced composites are studied to reveal strengthening mechanisms.

1.4 Thesis Overview

This thesis is organized as follows.

- **Chapter 1 Introduction**
  The research background is briefly introduced, the status of the research on mechanical properties of graphene and its composites is reviewed, and three key issues in this field are identified.
- **Chapter 2 Multiscale methods to investigate mechanical properties of graphene**
  Common multiscale methods of studying mechanical properties of graphene and its composites are reviewed, including molecular dynamics (MD) simulations and the first-principles calculations.
- **Chapter 3 Buckling of graphene monolayer under in-plane compression**
  Static buckling and wrinkle evolution in graphene monolayer under compression are examined. The related mechanisms are analyzed with the continuum plate theory and discontinuous effects.
- **Chapter 4 Dynamic rippling in graphene monolayer**
  Dynamic rippling in graphene monolayer is studied, and the related mechanism and potential applications in nanoelectromechanical systems (NEMS) are discussed.
- **Chapter 5 Defect-induced discontinuous effects in graphene nanoribbon under torsion loading**
  Defects in graphene are reviewed and discussed. A representative case including both static and dynamic out-of-plane deformation, the mechanical behavior of a graphene nanoribbon under torsion loading, is analyzed to reveal the effects of defects.
- **Chapter 6 Mechanical behaviors of graphene nanolayered composites**
  Mechanical properties of graphene nanolayered composites are studied under two representative loads (out-of-plane shock and in-plane shear), and the effects of graphene interfaces are discussed.
- **Chapter 7 Summary and future work**
  This chapter presents a summary of the work in this thesis, and perspectives about graphene and its composites in the field of nanomechanics.

References

1. Geim AK (2009) Science 324(5934):1530
2. Geim AK, Novoselov KS (2007) Nat Mater 6(3):183
3. Ezawa M (2006) Phys Rev B 73(4):045432

4. Balandin AA, Ghosh S, Bao W, Calizo I, Teweldebrhan D, Miao F, Lau CN (2008) Nano Lett 8(3):902
5. Liu M, Yin X, Ulin-Avila E, Geng B, Zentgraf T, Ju L, Wang F, Zhang X (2011) Nature 474(7349):64
6. Lee C, Wei X, Kysar JW, Hone J (2008) Science 321(5887):385
7. Stoller MD, Park S, Zhu Y, An J, Ruoff RS (2008) Nano Lett 8(10):3498
8. Stankovich S, Dikin DA, Dommett GH, Kohlhaas KM, Zimney EJ, Stach EA, Piner RD, Nguyen ST, Ruoff RS (2006) Nature 442(7100):282
9. Bunch JS, Verbridge SS, Alden JS, Van Der Zande AM, Parpia JM, Craighead HG, McEuen PL (2008) Nano Lett 8(8):2458
10. Wang Y, Shi Z, Huang Y, Ma Y, Wang C, Chen M, Chen Y (2009) J Phys Chem C 113(30):13103
11. Gui G, Li J, Zhong J (2008) Phys Rev B 78(7):075435
12. Bao W, Miao F, Chen Z, Zhang H, Jang W, Dames C, Lau CN (2009) Nat Nanotechnol 4(9):562
13. Konatham D, Striolo A (2008) Nano Lett 8(12):4630
14. Galiotis C, Frank O, Koukaras EN, Sfyris D (2015) Annu Rev Chem Biomol Eng 6:121
15. Huang X, Qi X, Boey F, Zhang H (2012) Chem Soc Rev 41(2):666
16. Ramanathan T, Abdala A, Stankovich S, Dikin D, Herrera-Alonso M, Piner R, Adamson D, Schniepp H, Chen X, Ruoff R et al (2008) Nat Nanotechnol 3(6):327
17. Gómez-Navarro C, Weitz RT, Bittner AM, Scolari M, Mews A, Burghard M, Kern K (2007) Nano Lett 7(11):3499
18. Wang S, Tambraparni M, Qiu J, Tipton J, Dean D (2009) Macromolecules 42(14):5251
19. Zhao X, Zhang Q, Chen D, Lu P (2010) Macromolecules 43(5):2357
20. Li C, Shi G (2012) Nanoscale 4(18):5549
21. Frank I, Tanenbaum DM, van der Zande AM, McEuen PL (2007) J Vac Sci Technol B Microelectron Nanometer Struct Process Meas Phenom 25(6):2558
22. Shokrieh MM, Rafiee R (2010) Mater Des 31(2):790
23. Huang Y, Wu J, Hwang KC (2006) Phys Rev B 74(24):245413
24. Liu F, Ming P, Li J (2007) Phys Rev B 76(6):064120
25. Van Lier G, Van Alsenoy C, Van Doren V, Geerlings P (2000) Chem Phys Lett 326(1–2):181
26. Konstantinova E, Dantas SO, Barone PM (2006) Phys Rev B 74(3):035417
27. Sánchez-Portal D, Artacho E, Soler JM, Rubio A, Ordejón P (1999) Phys Rev B 59(19):12678
28. Kudin KN, Scuseria GE, Yakobson BI (2001) Phys Rev B 64(23):235406
29. Zhao H, Min K, Aluru N (2009) Nano Lett 9(8):3012
30. Zhao H, Aluru N (2010) J Appl Phys 108(6):064321
31. Zhang P, Ma L, Fan F, Zeng Z, Peng C, Loya PE, Liu Z, Gong Y, Zhang J, Zhang X et al (2014) Nat Commun 5:3782
32. Lee JH, Loya PE, Lou J, Thomas EL (2014) Science 346(6213):1092
33. Zhang T, Li X, Kadkhodaei S, Gao H (2012) Nano Lett 12(9):4605
34. Hirano M, Shinjo K (1993) Wear 168(1–2):121
35. Shinjo K, Hirano M (1993) Surf Sci 283(1–3):473
36. Zheng Q, Jiang B, Liu S, Weng Y, Lu L, Xue Q, Zhu J, Jiang Q, Wang S, Peng L (2008) Phys Rev Lett 100(6):067205
37. Liu Z, Yang J, Grey F, Liu JZ, Liu Y, Wang Y, Yang Y, Cheng Y, Zheng Q (2012) Phys Rev Lett 108(20):205503
38. Zhang R, Ning Z, Zhang Y, Zheng Q, Chen Q, Xie H, Zhang Q, Qian W, Wei F (2013) Nat Nanotechnol 8(12):912
39. Wang W, Dai S, Li X, Yang J, Srolovitz DJ, Zheng Q (2015) Nat Commun 6:7853
40. Zan R, Muryn C, Bangert U, Mattocks P, Wincott P, Vaughan D, Li X, Colombo L, Ruoff RS, Hamilton B et al (2012) Nanoscale 4(10):3065
41. Thompson-Flagg RC, Moura MJ, Marder M (2009) Eur Lett 85(4):46002
42. He Y, Li H, Si P, Li Y, Yu H, Zhang X, Ding F, Liew KM, Liu X (2011) Appl Phys Lett 98(6):063101

43. Fasolino A, Los J, Katsnelson MI (2007) Nat Mater 6(11):858
44. Shen HS, Xu YM, Zhang CL (2013) Appl Phys Lett 102(13):131905
45. Guinea F, Horovitz B, Le Doussal P (2008) Phys Rev B 77(20):205421
46. Morozov S, Novoselov K, Katsnelson M, Schedin F, Elias D, Jaszczak JA, Geim A (2008) Phys Rev Lett 100(1):016602
47. Neto AC, Guinea F, Peres NM, Novoselov KS, Geim AK (2009) Rev Mod Phys 81(1):109
48. De Parga AV, Calleja F, Borca B, Passeggi M Jr, Hinarejos J, Guinea F, Miranda R (2008) Phys Rev Lett 100(5):056807
49. Kim KS, Zhao Y, Jang H, Lee SY, Kim JM, Kim KS, Ahn JH, Kim P, Choi JY, Hong BH (2009) Nature 457(7230):706
50. San-José P, González J, Guinea F (2011) Phys Rev Lett 106(4):045502
51. Singh AK, Hennig RG (2013) Phys Rev B 87(9):094112
52. Osvath Z, Lefloch F, Bouchiat V, Chapelier C (2013) Nanoscale 5(22):10996
53. Wang C, Liu Y, Lan L, Tan H (2013) Nanoscale 5(10):4454
54. Paronyan TM, Pigos EM, Chen G, Harutyunyan AR (2011) ACS Nano 5(12):9619
55. Gao L, Guest JR, Guisinger NP (2010) Nano Lett 10(9):3512
56. Locatelli A, Wang C, Africh C, Stojic N, Mentes TO, Comelli G, Binggeli N (2013) ACS Nano 7(8):6955
57. Mao Y, Wang WL, Wei D, Kaxiras E, Sodroski JG (2011) ACS Nano 5(2):1395
58. Lahiri D, Das S, Choi W, Agarwal A (2012) ACS Nano 6(5):3992
59. Koch S, Stradi D, Gnecco E, Barja S, Kawai S, Diaz C, Alcami M, Martin F, Vazquez de Parga AL, Miranda R et al (2013) ACS Nano 7(4):2927
60. Kim SY, Park HS (2011) J Appl Phys 110(5):054324
61. Garcia-Sanchez D, van der Zande AM, Paulo AS, Lassagne B, McEuen PL, Bachtold A (2008) Nano Lett 8(5):1399
62. Lovat G, Burghignoli P, Araneo R (2013) IEEE Trans Electromagn Compat 55(2):328
63. Yazyev OV, Helm L (2007) Phys Rev B 75(12):125408
64. Chen JH, Li L, Cullen WG, Williams ED, Fuhrer MS (2011) Nat Phys 7(7):535
65. Krasheninnikov A, Lehtinen P, Foster AS, Pyykkö P, Nieminen RM (2009) Phys Rev Lett 102(12):126807
66. Botello-Méndez AR, Declerck X, Terrones M, Terrones H, Charlier JC (2011) Nanoscale 3(7):2868
67. Ferreira A, Xu X, Tan CL, Bae SK, Peres N, Hong BH, Özyilmaz B, Neto AC (2011) Eur Lett 94(2):28003
68. Bell DC, Lemme MC, Stern LA, Williams JR, Marcus CM (2009) Nanotechnology 20(45):455301
69. Krasheninnikov A, Banhart F (2007) Nat Mater 6(10):723
70. Fischbein MD, Drndić M (2008) Appl Phys Lett 93(11):113107
71. Lemme MC, Bell DC, Williams JR, Stern LA, Baugher BW, Jarillo-Herrero P, Marcus CM (2009) ACS Nano 3(9):2674
72. Goverapet Srinivasan S, van Duin AC (2011) J Phys Chem A 115(46):13269
73. Wang H, Wang Q, Cheng Y, Li K, Yao Y, Zhang Q, Dong C, Wang P, Schwingenschlogl U, Yang W et al (2011) Nano Lett 12(1):141
74. Lehtinen O, Kotakoski J, Krasheninnikov A, Tolvanen A, Nordlund K, Keinonen J (2010) Phys Rev B 81(15):153401
75. Åhlgren E, Kotakoski J, Krasheninnikov A (2011) Phys Rev B 83(11):115424
76. Jack R, Sen D, Buehler MJ (2010) J Comput Theor Nanosci 7(2):354
77. Long X, Zhao F, Liu H, Huang J, Lin Y, Zhu J, Luo S (2015) J Phys Chem C 119(13):7453
78. Kim K, Lee Z, Regan W, Kisielowski C, Crommie M, Zettl A (2011) ACS Nano 5(3):2142
79. Grantab R, Shenoy VB, Ruoff RS (2010) Science 330(6006):946
80. Wei Y, Wu J, Yin H, Shi X, Yang R, Dresselhaus M (2012) Nat Mater 11(9):759
81. Yin H, Qi HJ, Fan F, Zhu T, Wang B, Wei Y (2015) Nano Lett 15(3):1918
82. Zhang J, Zhao J, Lu J (2012) ACS Nano 6(3):2704
83. Neek-Amal M, Peeters F (2012) Appl Phys Lett 100(10):101905

84. Zhang T, Li X, Gao H (2014) J Mech Phys Solids 67:2
85. Qi X, Pu KY, Li H, Zhou X, Wu S, Fan QL, Liu B, Boey F, Huang W, Zhang H (2010) Angew Chem Int Ed 49(49):9426
86. He Q, Sudibya HG, Yin Z, Wu S, Li H, Boey F, Huang W, Chen P, Zhang H (2010) ACS Nano 4(6):3201
87. Li B, Cao X, Ong HG, Cheah JW, Zhou X, Yin Z, Li H, Wang J, Boey F, Huang W et al (2010) Adv Mater 22(28):3058
88. Qi X, Pu KY, Zhou X, Li H, Liu B, Boey F, Huang W, Zhang H (2010) Small 6(5):663
89. Zhu Y, Murali S, Cai W, Li X, Suk JW, Potts JR, Ruoff RS (2010) Adv Mater 22(35):3906
90. Cong HP, Ren XC, Wang P, Yu SH (2012) ACS Nano 6(3):2693
91. Cong HP, Ren XC, Wang P, Yu SH (2012) Sci Rep 2:613
92. Xu WP, Zhang LC, Li JP, Lu Y, Li HH, Ma YN, Wang WD, Yu SH (2011) J Mater Chem 21(12):4593
93. Cong HP, Ren XC, Wang P, Yu SH (2013) Energy Environ Sci 6(4):1185
94. Rafiee MA, Lu W, Thomas AV, Zandiatashbar A, Rafiee J, Tour JM, Koratkar NA (2010) ACS Nano 4(12):7415
95. Walker LS, Marotto VR, Rafiee MA, Koratkar N, Corral EL (2011) ACS Nano 5(4):3182
96. Tian M, Qu L, Zhang X, Zhang K, Zhu S, Guo X, Han G, Tang X, Sun Y (2014) Carbohydr Polym 111:456
97. Bartolucci SF, Paras J, Rafiee MA, Rafiee J, Lee S, Kapoor D, Koratkar N (2011) Mater Sci Eng A 528(27):7933
98. Gong L, Kinloch IA, Young RJ, Riaz I, Jalil R, Novoselov KS (2010) Adv Mater 22(24):2694
99. Gong C, Lee G, Shan B, Vogel EM, Wallace RM, Cho K (2010) J Appl Phys 108(12):123711
100. Kuilla T, Bhadra S, Yao D, Kim NH, Bose S, Lee JH (2010) Prog Polym Sci 35(11):1350
101. Geim AK, Grigorieva IV (2013) Nature 499(7459):419
102. Kim Y, Lee J, Yeom MS, Shin JW, Kim H, Cui Y, Kysar JW, Hone J, Jung Y, Jeon S et al (2013) Nat Commun 4:3114
103. Goli P, Ning H, Li X, Lu CY, Novoselov KS, Balandin AA (2014) Nano Lett 14(3):1497
104. Zhang D, Zhan Z (2016) J Alloy Compd 654:226
105. Hong D, Bae K, Yoo S, Kang K, Jang B, Kim J, Kim S, Jeon S, Nam Y, Kim YG et al (2014) Macromol Biosci 14(3):314
106. Choi BG, Yang M, Hong WH, Choi JW, Huh YS (2012) ACS Nano 6(5):4020
107. Huang L, Yi N, Wu Y, Zhang Y, Zhang Q, Huang Y, Ma Y, Chen Y (2013) Adv Mater 25(15):2224
108. Hoagland RG, Kurtz RJ, Henager C Jr (2004) Scr Mater 50(6):775
109. Liu J, Yan H, Jiang K (2013) Ceram Int 39(6):6215
110. Roman RE, Cranford SW (2014) Adv Eng Mater 16(7):862
111. Fang M, Wang K, Lu H, Yang Y, Nutt S (2009) J Mater Chem 19(38):7098
112. Loomis J, King B, Panchapakesan B (2012) Appl Phys Lett 100(7):073108
113. Terrones M, Botello-Méndez AR, Campos-Delgado J, López-Urías F, Vega-Cantú YI, Rodríguez-Macías FJ, Elías AL, Munoz-Sandoval E, Cano-Márquez AG, Charlier JC et al (2010) Nano Today 5(4):351
114. Wang J, Hoagland R, Hirth J, Misra A (2008) Acta Mater 56(13):3109

Chapter 2
Multiscale Methods to Investigate Mechanical Properties of Graphene

This chapter begins with a brief review of the common methods for studying the mechanical properties of graphene and its composites, and then introduces the molecular dynamics method and the first-principle calculations used in this thesis. A method of converting physical quantities from molecular simulations to continuum models is explained. Some C++ codes for simulation and modeling in this thesis are introduced.

2.1 Introduction

Compared with common three-dimensional (3D) materials, the methods to study the mechanical properties of graphene are unique due to its monoatomic thickness. During these explorations, innovative methods can be obtained, which is of great significance for further research on low-dimensional nanomaterials. In this section, the common methods of studying the mechanical properties of graphene and its composites are reviewed from three aspects: experiment, atomistic simulation, and theoretical analysis.

- **Experiment**
  Experiment is undoubtedly the fundamental method. In 2015, Galiotis et al. provide a review of current experimental methods to characterize the mechanical properties of graphene [1]: direct methods and Raman spectroscopy are the two main experimental methods. For the former, the direct methods refer to directly measure the force and deformation of the graphene under loading such as nanoindentation [2, 3]; for the latter, the mechanical properties of graphene are obtained indirectly by Raman spectroscopy [4]. The mechanical parameters of graphene obtained by direct methods are highly influenced by the experimental conditions and the quality of graphene [5]: in order to obtain reliable experimental results, a large number of repeated experiments are necessary. Recently, with the

continuous improvement of fabrication technology, the quality of graphene is improved, and the direct methods become more and more mature [6]. For the Raman spectroscopy, its capabilities go far beyond the commonly used assessment of quality and number of layers, which is only suitable for studying the mechanical properties of pristine graphene and functionalized graphene [7].

- **Atomistic Simulation**

  In recent years, atomistic simulation has become an important method to study the mechanical properties of graphene and its composites. The most common simulation methods include the first-principle calculation based on density functional theory (DFT) [8], molecular mechanics method [9], Monte Carlo (MC) method [10], molecular dynamics (MD) method [11]. The DFT simulation is based on the Schrödinger equations and calculating the interaction between atoms based on the electronic plane wave function, which is the most accurate method of atomistic simulation methods. In general, pseudopotential is used for simplifying the description of complex systems in DFT simulation [12]. Even using pseudopotential instead of full-potential, the computational cost of DFT simulation is still proportional to the cube of the number of atoms [13]. Although the linear-scaling DFT simulation methods have been improved in recent years [14], the upper limitation of the number of atoms for DFT simulations is generally considered as ∼1,000 [15]. Therefore, this method is not proper to simulate graphene at large scale, which is generally used to study the electrical properties (e.g., band structures and electron transport). The molecular mechanics, MC and MD methods are all based on the the description of the atomic interactions based on Newton's laws. For the molecular mechanics method, the potential energy of the whole system is calculated as a function of the coordinates of atoms, and this method uses an appropriate algorithm (e.g., steepest descent) to find the molecular structure of a local energy minimum, which refers to the static equilibrium. The static and quasi-static mechanical properties of graphene can be studied by the molecular mechanics method, while the system temperature is 0 K due to the regardless of the kinetic energy. Similarly, the MC method also obtains the equilibrium state by solving the minimum value of the system energy. Although the MC method can define a temperature of the system by the Boltzmann constant, its "dynamic" process refers to the adjustment of atomic position, which is not a realistic movement of atoms. For the MD method, the trajectories of atoms are determined by numerically solving Newton's equations of motion, where interactions between the atoms and their potential energies are generally calculated using interatomic potentials or molecular mechanics force fields. The kinetic energy and potential energy are both calculated in MD simulations. Therefore, the temperature and the realistic dynamic process of the atoms can be characterized simultaneously. However, since the time step used in MD simulations is generally at femtosecond scale, the simulated strain rate is several orders of magnitude higher than the experimental value, and this effect should be considered in studying the mechanical behaviors of graphene and its composites.

2.1 Introduction

- **Theoretical Analysis**
  The mechanical behaviors of graphene and its composites are generally characterized by equivalent models based on continuum mechanics [16]. As mentioned in Chap. 1, different continuum mechanical models are applicable for different mechanical behaviors of graphene [17, 18]. A common equivalent continuum mechanical model of graphene is plate [19], and the interfacial interactions induced by graphene are approximated as spring or viscous forces [20, 21]. However, for the plate model, the thickness of graphene is not well defined. Different values of equivalent thickness are applied to characterize the in-plane tension [16], wrinkling [22] and out-of-plane vibration [23], etc. Recently, based on the atomic interaction, the quasi-continuous mechanics method has also been developed [24, 25]: the Cauchy–Born model is used to convert the atomic potential energy to the strain energy term in continuum mechanical models [26]. However, the discontinuous effects mentioned in Chap. 1 can not be characterized by these continuum models (e.g., the chirality-dependent mechanical behaviors of graphene), which will be elaborated in following chapters.

To resolve the three important issues raised in Chap. 1, in this thesis, the mechanical behaviors of graphene and graphene composites are simulated by MD method, the mechanisms of discontinuous effects are revealed by DFT simulations, and the mechanical properties are analyzed by theoretical mechanical models, respectively.

2.2 MD and DFT Simulation Methods

The details of MD and DFT simulation methods used in this thesis are elaborated in this section.

2.2.1 MD Simulation

There are four steps for MD simulations (Fig. 2.1).

- **Modeling Initial Atomistic Configuration**
  The first step of MD simulation is modeling an initial atomistic configuration. Initial configurations can be obtained from experimental data or quantum chemical calculations with a reasonable system energy. Unreasonable atomistic configuration may result unreasonable system energy (e.g., zero distance between two atoms results in infinity value of force). If necessary, energy minimization can be performed to reduce initial system energy. For example, the grain boundaries in graphene monolayer are studied in this thesis, which are complex for modeling. C++ codes are developed in this thesis (Appendix A) to model grain boundaries in graphene monolayer with mismatch angles of 5.5°, 13.2° and 21.7° (Fig. 2.2), and

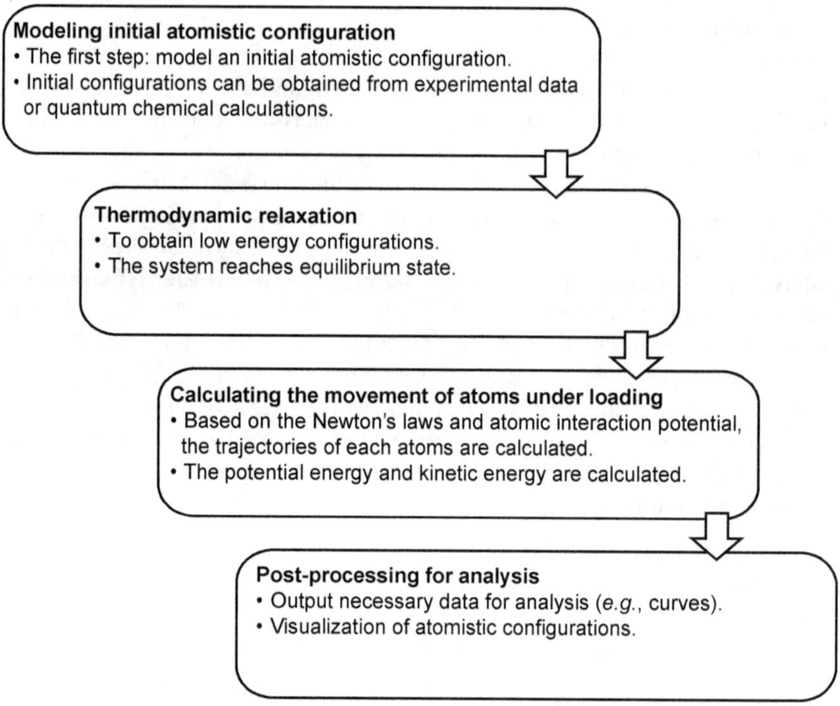

Modeling initial atomistic configuration
- The first step: model an initial atomistic configuration.
- Initial configurations can be obtained from experimental data or quantum chemical calculations.

⇓

Thermodynamic relaxation
- To obtain low energy configurations.
- The system reaches equilibrium state.

⇓

Calculating the movement of atoms under loading
- Based on the Newton's laws and atomic interaction potential, the trajectories of each atoms are calculated.
- The potential energy and kinetic energy are calculated.

⇓

Post-processing for analysis
- Output necessary data for analysis (*e.g.*, curves).
- Visualization of atomistic configurations.

Fig. 2.1 Four steps for MD simulations

Fig. 2.2 Grain boundaries in graphene monolayer with different mismatch angles of 5.5°, 13.2° and 21.7°

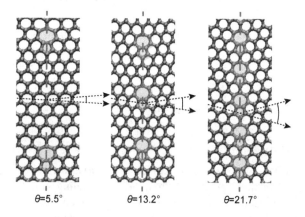

$\theta=5.5°$ $\theta=13.2°$ $\theta=21.7°$

the energy minimization is performed for these models before thermodynamics relaxation.

- **Thermodynamics Relaxation**
  In general, the initial atomistic configurations do not reach equilibrium state, and the thermodynamics relaxation is necessary. The thermodynamics relaxation is actually a process of energy redistribution of the initial configuration by changing the positions of atoms, which leads to an equilibrium state of system. Random velocities according to Boltzmann distribution are applied to atoms to achieve desired temperature. Temperature and pressure controls for different ensembles adjust atoms with spontaneous energy corrections. The system can reach equilibrium state with long enough time (large enough simulation steps). After thermodynamics relaxation, the atomistic configuration reaches equilibrium state and becomes realistic, which is the premise of following simulations.

- **Calculating the Movement of Atoms Under Loading**
  After thermodynamic relaxation, the system reaches equilibrium state, and then desired loading is applied. The displacements of atoms are calculated by Newton's law. At time step n, the force between atom i and j is determined by the derivative of their interaction potential with respect to distance:

$$f_{i,j}^n = -\frac{\partial E}{\partial r_{i,j}^n}. \tag{2.1}$$

Assuming that there are M atoms in the system. The net force of atom i is calculated as the vector summation of other atoms excluding atom i itself:

$$F_j^n = \sum_{j \neq i}^{M} f_{i,j}^n. \tag{2.2}$$

According to Newton's second law, at the next time step $n + 1$, the displacement r_i^{n+1} and velocity v_i^{n+1} of atom i is given as:

$$\begin{cases} v_i^{n+1} = v_i^n + F_i^n \Delta t / m_i \\ r_i^{n+1} = r_i^n + v_i^n \Delta t \end{cases}, \tag{2.3}$$

where Δt is the time step, and m_i is the mass of atom i. Through calculating the forces in the system, the displacements and velocities of all atoms can be obtained every time step. When the distance between two atoms is overly large, the atomic interaction between them is very weak and negligible. To decrease computing cost, Eq. (2.2) can be approximately given with neglecting the weak forces:

$$F_j^n = \sum_{j \neq i}^{M} f_{i,j}^n \quad \text{at} \quad \| r_{i,j}^n \| < r_{\text{cutoff}}, \tag{2.4}$$

where r_{cutoff} is referred as "cutoff distance". In general, r_{cutoff} is about 10 times of the equilibrium distance between two atoms.
- **Post-Processing for Analysis**
Post-processing should be performed for analysis after simulations. The origin data from MD simulations consists of atomic displacements and velocities, and all other physical quantities are derived from these two variables. Recently, most MD simulation software provides the functions to directly output some derived physical quantities (e.g., temperature, potential energy and kinetic energy). However, some physical quantities cannot be obtained directly, which should be converted from the basic by specified methods.

In this thesis, all MD simulations are performed by open source codes Large-scale Atomic/Molecular Massively Parallel Simulator (LAMMPS) [27]. The detail of MD simulations will be discussed in following chapters.

2.2.2 DFT Simulation

According to the Schrödinger equations, the atomic interaction is a superposition of electron wave functions. The main goal of density functional theory (DFT) is to replace the wave function with electron density as the fundamental research quantities. This method is used in physics, chemistry and materials science to investigate the electronic structure (principally the ground state) of many-body systems, in particular atoms, molecules, and the condensed phases.

DFT can be traced back to the Thomas–Fermi model developed by Thomas and Fermi in 1920s [28, 29]: the atomic energy is calculated as a functional of electron density with considering the classical expression of the nucleus–electron and electron–electron interactions (both of which can be expressed by electron density). However, the Thomas–Fermi model does not consider the atomic exchange energy in Hartree–Fock theory. Although Dirac added an exchange energy functional term to this model, the Thomas–Fermi–Dirac theory is still very inaccurate. The largest error comes from the representation of kinetic energy, the exchange energy, and the complete neglect of the electron-related effects.

Although the Thomas–Fermi model proposes a basic idea of characterizing the properties of atomistic system using electron density, it is still difficult to be used as an applicable method due to the errors mentioned above. These errors are not resolved until the Hohenberg–Kohn theorem (H-K theorem) is put forward [30]. The H-K theorem defines an energy functional for the system and proves that the correct ground state electron density minimizes this energy functional. The H-K theorem is further developed by Kohn and Sham to produce Kohn–Sham DFT (KS DFT), which makes DFT becomes an applicable calculation method [31].

In recent years, with the development of computer science, the DFT simulation method has been greatly improved, and there are many open source codes to be used. In this thesis, the open source software OpenMX [32] and Quantum-ESPRESSO [33] are used to calculate the electron density distribution and electron transport characteristics of graphene and its composites. The detail of the related simulations is discussed in following chapters.

2.3 Conversion of Physical Quantities from MD Simulations to Continuum Models

According to the previous introduction, two basic quantities obtained by molecular dynamics simulation are the atomic positions and velocities. To compare with macroscopic mechanical theory, it is necessary to extract the macroscopic physical quantities from the molecular simulation results. This section will discuss the conversion of two representative physical quantities from atomistic to continuum.

- **Temperature**

  In MD simulations, temperature T is proportional to atomic kinetic energy, and the relation is given as:

$$\sum_i^N \frac{1}{2} m_i v_i^2 = \frac{\dim}{2} NkT, \qquad (2.5)$$

  where N is the number of atoms in system, m_i and v_i are the mass and velocity of atom i, dim is the dimension of system (dim $=2$ for 2D, dim $=3$ for 3D), and k is the Boltzmann constant.

  In most cases, the system temperature can be calculated accurately by Eq. (2.5). However, this equation should be corrected when the center of mass velocity of system cannot be neglected. A well-known example is the "Flying Ice" effect [34]: using molecular dynamics to simulate a high-speed gliding ice cube, the kinetic energy of this ice cube can be extremely high, which leads to a system temperature over water vaporization temperature according to Eq. (2.5). Actually, common sense tells us that the flying ice cube is still cold and exists as a solid. Therefore, in this situation, Eq. (2.5) should be corrected as:

$$\sum_i^N \frac{1}{2} m_i (v_i - v_c)^2 = \frac{\dim}{2} NkT, \qquad (2.6)$$

  where v_c is the center of mass velocity of system. In the following chapters, this effect should be considered in the calculation of the temperature for vibrational graphene monolayer.

- **Stress**

  Stress is a concept in continuum mechanics:

$$\sigma_{ij} = \lim_{\Delta A_i \to 0} \frac{\Delta F_j}{\Delta A_i}, \qquad (2.7)$$

where σ_{ij} is stress, ΔF_j is the force in direction j, ΔA_i is the loading area in direction i. Continuous mass is the basic assumption of this definition. However, in MD simulations, the mass of atoms is discrete. To compare with continuum mechanical models, it is necessary to extract the equivalent stress form MD simulations.

According to the introduction of MD simulations, the force on each atom can be obtained (Eq. 2.4). However, since atoms are considered as particles, the area A_i cannot be well defined in Eq. (2.7). To avoid misunderstanding discussion of A_i, the Virial stress is widely used in MD simulations:

$$\tau_{ij} = \frac{1}{\Omega} \sum_{k \in \Omega} \left(-m^{(k)}(u_i^{(k)} - \bar{u}_i)(u_j^{(k)} - \bar{u}_j) + \frac{1}{2} \sum_{\ell \in \Omega} (x_i^{(\ell)} - x_i^{(k)}) f_j^{(k\ell)} \right) \quad (2.8)$$

where k and ℓ are atoms in the domain, Ω is the volume of the domain (Virial volume), $m^{(k)}$ is the mass of atom k, $u_i^{(k)}$ is the ith component of the velocity of atom k, $\bar{u}_j$ is the jth component of the average velocity of atoms in the volume, $x_i^{(k)}$ is the ith component of the position of atom k, and $f_i^{(k\ell)}$ is the ith component of the force applied on atom k by atom ℓ, respectively.

In MD simulations, the Virial volume can be calculated by the Voronoi methods: the polyhedron composed of symmetry planes between an atom and all its neighboring atoms is the Virial cell. The Virial volume of internal atoms is reasonable, while that of surface atoms is obviously overestimated (Fig. 2.3). In addition, since the thermodynamic state of a single atom is fluctuating rapidly at femtosecond scale, the stress of a single atom is meaningless due to huge error. In this situation, the

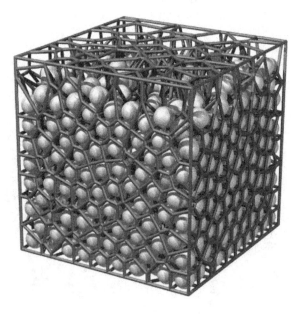

Fig. 2.3 Virial cells calculated by Voronoi method

error can be reduced by calculating a statistical results of atoms in a large Ω. In this thesis, the MD results show that, when $\Omega > 5$ nm^3, the Virial stress is accurate. In the following chapters, to resolve the stress of graphene composites under shock loading, a series of C++ codes are developed, which have been posted in Appendix B.

2.4 Chapter Summary

In this chapter, the common methods for studying the mechanical properties of graphene and its composites are reviewed from three aspects: experiment, atomistic simulation and theoretical analysis. The detail of MD and DFT simulation method used in the following chapters is introduced, and the conversion of physical quantities from MD simulations to continuum mechanical models is discussed.

References

1. Galiotis C, Frank O, Koukaras EN, Sfyris D (2015) Annu Rev Chem Biomol Eng 6:121
2. Lee C, Wei X, Kysar JW, Hone J (2008) Science 321(5887):385
3. Huang PY, Ruiz-Vargas CS, van der Zande AM, Whitney WS, Levendorf MP, Kevek JW, Garg S, Alden JS, Hustedt CJ, Zhu Y et al (2011) Nature 469(7330):389
4. Ferrari AC, Basko DM (2013) Nat Nanotechnol 8(4):235
5. Huang M, Pascal TA, Kim H, Goddard WA III, Greer JR (2011) Nano Lett 11(3):1241
6. Wu T, Zhang X, Yuan Q, Xue J, Lu G, Liu Z, Wang H, Wang H, Ding F, Yu Q et al (2016) Nat Mater 15(1):43
7. Frank O, Tsoukleri G, Parthenios J, Papagelis K, Riaz I, Jalil R, Novoselov KS, Galiotis C (2010) ACS Nano 4(6):3131
8. Chester M (2012) Primer of quantum mechanics. Courier Corporation
9. Burkert U, Allinger N (1982) Molecular mechanics. American Chemical Society, Washington, DC
10. Baumgärtner A, Burkitt A, Ceperley D, De Raedt H, Ferrenberg A, Heermann D, Herrmann H, Landau D, Levesque D, von der Linden W, et al (2012) The Monte Carlo method in condensed matter physics, vol 71. Springer Science & Business Media
11. Alder BJ, Wainwright TE (1959) J Chem Phys 31(2):459
12. Hellmann H (1935) J Chem Phys 3(1):61
13. Sholl D, Steckel JA (2011) Density functional theory: a practical introduction. Wiley
14. Haynes P, Skylaris CK, Mostofi A, Payne M (2004) In: APS Meeting Abstracts
15. March N (1982) J Chem Phys 86(12):2262
16. Shokrieh MM, Rafiee R (2010) Mater Des 31(2):790
17. Kireitseu M, Kompis V, Altenbach H, Bochkareva V, Hui D, Eremeev S (2005) Fullerenes. Nanotub Carbon Nanostructures 13(4):313
18. Odegard GM, Gates TS, Nicholson LM, Wise KE (2002) Compos Sci Technol 62(14):1869
19. Wang Q (2005) J Appl Phys 98(12):124301
20. Kitipornchai S, He X, Liew K (2005) Phys Rev B 72(7):075443
21. Lu W, Wu J, Song J, Hwang K, Jiang L, Huang Y (2008) Comput Methods Appl Mech Eng 197(41–42):3261
22. Wang C, Mylvaganam K, Zhang L (2009) Phys Rev B 80(15):155445

23. Kim SY, Park HS (2011) J Appl Phys 110(5):054324
24. Wang X, Guo X (2013) J Comput Theor Nanosci 10(1):154
25. Arroyo M, Belytschko T (2004) Phys Rev B 69(11):115415
26. Zhang P, Jiang H, Huang Y, Geubelle P, Hwang K (2004) J Mech Phys Solids 52(5):977
27. Plimpton S (1995) J Comput Phys 117(1):1
28. Thomas LH (1927) In: Mathematical Proceedings of the Cambridge Philosophical Society, vol 23. Cambridge University Press, pp 542–548
29. Fermi E (1927) Rend Accad Naz Lincei 6(602–607):32
30. Hohenberg P, Kohn W (1964) Phys Rev B 136(3):B864
31. Kohn W, Sham LJ (1965) Phys Rev A 140(4):A1133
32. Boker S, Neale M, Maes H, Wilde M, Spiegel M, Brick T, Spies J, Estabrook R, Kenny S, Bates T et al (2011) Psychometrika 76(2):306
33. Giannozzi P, Baroni S, Bonini N, Calandra M, Car R, Cavazzoni C, Ceresoli D, Chiarotti GL, Cococcioni M, Dabo I et al (2009) J Phys Condens Matter 21(39):395502
34. Harvey SC, Tan RKZ, Cheatham TE III (1998) J Comput Chem 19(7):726

Chapter 3
Buckling of Graphene Monolayer Under In-Plane Compression

For graphene monolayer, the out-of-plane induced by in-plane loading is buckling, which is one of the basic out-of-plane deformations raised in Chap. 1. The static and dynamic buckling of graphene monolayer is studied by theoretical mechanical models and numerous simulations in this chapter.

3.1 Introduction

The interest of graphene increases due to its exceptional electrical, mechanical and optical properties [1]. Controllable mechanical behaviors of graphene under various stress states lead to applications such as sensors [2, 3], superconductors [4, 5], and supercapacitors [6] in nanoelectromechanical systems (NEMS) [7]. Buckling is one of the important mechanical behaviors of graphene [8]. Buckling-induced wrinkles and ripples in graphene can effectively influence the gauge field [9, 10], electronic structure [11, 12], Raman spectroscopy [13, 14], electronic transport [15, 16] and carrier mobilities [17], etc. In addition, the growing interest in the exploitation of graphene as nanoreinforcement [18, 19] makes it important to understand the buckling behavior of graphene [20]. Understanding the underlying mechanism of buckling in graphene monolayer is important for the design of graphene and its composites [21], which is also one of the important issues raised in Chap. 1: static and dynamic out-of-plane deformation of graphene under in-plane loading.

Wrinkles and ripples are generally resulted by interfacial instability under in-plane compression [22], which are ubiquitous in the epitaxial graphene on metal surfaces [23, 24]. Studying the mechanical behaviors under in-plane compression is an effective approach to reveal the underlying mechanism of buckling in graphene monolayer [25, 26]. Continuum equivalent plate models [27] have been used to study the mechanical properties of graphene monolayer under uniaxial [28], biaxial [29] and hydrostatic stress [14]. In addition, although the Young's modulus and bending stiffness may be affected by chirality under large deformation [30, 31], the buckling in graphene monolayer is considered isotropic in most continuum mechanical models [26]. This effect is crucial in the design of graphene-based nanodevices. Moreover, with the development of terahertz nanosensors [32], the dynamic evolution of wrinkles and ripples in graphene monolayer is also important for designing nanodevices with high frequency response [33–35].

In this chapter, first-principles calculations, molecular dynamics (MD) simulations, continuum mechanical theory, and discontinuous analysis are used to investigate the static and dynamic buckling behaviors of graphene monolayer under in-plane compression. The results show that, the buckling of graphene monolayer is remarkably chirality- and size-dependent. The local aspect ratio (the ratio of width to length) of wrinkles in graphene monolayer is an important parameter to characterize the static buckling behavior, which also highly influences the dynamic evolution of wrinkles. As the size of wrinkles increases from nanoscale to macroscale, the discontinuous effect, the plate-bending stiffness, and the membrane properties have a dominant influence on the buckling behavior of graphene monolayer, successively. Correspondingly, the preferred growth direction of buckling-induced wrinkles in graphene monolayer behaves as zigzag-along, armchair-along and isotropic. Through studying the buckling of graphene monolayer, its basic static and dynamic out-of-plane behaviors under in-plane compression can be revealed, which can also lead to an improved fundamental understanding of the buckling of graphene-based devices in following chapters.

3.2 Simulation Details

3.2.1 MD Simulation

MD simulations are performed using open codes Large-scale Atomic/Molecular Massively Parallel Simulator (LAMMPS) [36]. To compare with the results of energy minimization (the temperature is 0 K) [37], before buckling simulations, the suspended graphene sheet is relaxed for 40 ps with canonical (NVT) ensemble at 0 K. The damping time constant is set to 0.1 ps, and the time step of all simulations is set to 1.0 fs. For finite models, the graphene sheet is placed in the central domain of a fixed simulation box with a vacuum space of 4 nm. For infinite models, periodic boundary conditions are applied in the basal plane of graphene without vacuum space. The interactions between carbon atoms are described by the adaptive intermolecular reactive empirical bond order (AIREBO) potential [38]. Stable buckling configurations of graphene monolayer are obtained via energy minimization simulations. The thickness of vacuum space is set to 40 nm along the out-of-plane direction. The pre-stress is applied by shrinking the lattice of graphene. Visualization is performed using free software AtomEye [39].

3.2 Simulation Details

3.2.2 DFT Simulation

DFT simulations with local density approximation (LDA) are used to study the discontinuous effects. The plane-wave basis with ultrasoft pseudopotential proposed by Vanderbilt is used for C atoms [40]. The cutoff energy is 240 eV, and the convergence in energy and force is set to 1×10^{-5} eV and 0.03 eVÅ^{-1} [41]. For infinite models, the dimensions of the suppercell should be compatible with the lattice of graphene (2.46 Å along zigzag direction, 4.26 Å along armchair direction). For finite models, the thickness of vacuum space along the in-plane and out-of-plane direction is 20 Å and 40 Å, respectively. To reduce computational cost of DFT simulations, the initial atomistic configurations of static wrinkles are imported from the results of energy minimization simulations.

3.3 The Buckling of Graphene Monolayer Under Hydrostatic Stress

In this section, MD simulations are performed to study the buckling of a circular graphene monolayer with diameter of 60 nm under hydrostatic stress. Since the system temperature is 0 K, out-of-plane perturbation should be applied to induce in-plane instability. Here, two typical out-of-plane perturbations, atom bombardment [42] and structural perturbation [43], are used. For the atom bombardment simulations, an iron atom with incident energy of 20 eV penetrates through the center of pre-strained graphene without inducing defects. For structural perturbation simulations, a small out-of-plane deflection of 0.5 Å is applied to the carbon atoms in the central domain of graphene monolayer within a diameter of 1 nm.

For the atom bombardment simulation, a six-petal-like pattern of wrinkles forms in the central domain of graphene sheet, whose petals are along zigzag direction (Fig. 3.1a). For the structural perturbation simulation (Fig. 3.1b), sprout-like pattern (region B) appears at the outside edge, and a hexagram-like pattern (region A) appears in the central domain. The hexagon ripples between region A and B are caused by the transverse waves (out-of-plane vibration) but not buckling, which will be discussed in following chapters. In addition, the buckling is zigzag-along in region A and armchair-along in region B, respectively. These two simulations demonstrate two issues below:

- The buckling in graphene monolayer is anisotropic.
- The buckling direction is size- and chirality-dependent.

These two issues have rarely been considered in most continuum mechanical models of buckling in graphene monolayer [44].

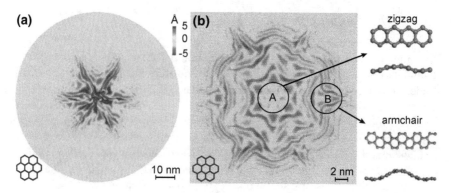

Fig. 3.1 The partial snapshots of wrinkles in graphene induced by **a** atom bombardment and **b** structural perturbation. Colorbar: out-of-plane displacement. Layout of graphene is inserted at the left bottom. The detail of region A and B shows that zigzag-along and armchair-along buckling occurs, respectively

3.4 Discontinuous Effects in the Buckling of Graphene Monolayer

A one-dimensional (1D) compressed buckling model of graphene monolayer is used To clarify the underlying mechanism of these two issues (Fig. 3.2). The parallel edges of the graphene sheet along y-axis are unsupported. In the classical Euler regime, the critical buckling strain ε_c for this model is given as [45]:

$$\varepsilon_c = \frac{k}{w^2} \frac{D\pi^2}{C}, \qquad (3.1)$$

where w and l are the width and length of the graphene sheet, $k = (mw/l + l/mw)^2$ is a geometric term, m is the number of wrinkles' half waves, and D and C are the flexural and tension rigidities, respectively.

According to Eq. (3.1), the aspect ratio $\alpha = w/l$ highly affects the critical buckling strain ε_c. Actually, the local aspect ratio varies with the growth of wrinkles (Fig. 3.1). Therefore, the effect of α should be jointly considered with strain ε to study the buckling properties of graphene monolayer.

For MD and energy minimization simulations, periodic boundary conditions are applied along width direction to simulate infinite w. According to Eq. (3.1), when the width of graphene sheet is infinite ($w \to \infty$), the critical buckling strain $\varepsilon_{c,p}$ is the minimum:

$$\varepsilon_{c,p} = \frac{1}{l^2} \frac{D\pi^2}{C}. \qquad (3.2)$$

l is the the only geometric factor affecting $\varepsilon_{c,p}$. The value of D/C is generally considered as isotropic [30, 43]. Therefore, there should not be chiral difference for $\varepsilon_{c,p}$ theoretically.

3.4 Discontinuous Effects in the Buckling of Graphene Monolayer

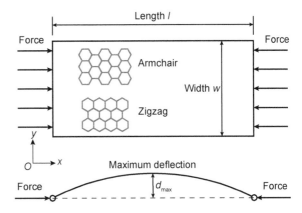

Fig. 3.2 The schematic of 1D compressed buckling model of graphene with unsupported parallel edges. Both the armchair- and zigzag-along compression are studied

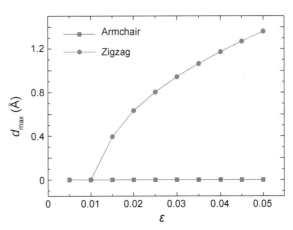

Fig. 3.3 The ε–d_{max} curves for armchair- and zigzag-along compression simulations

However, the results of energy minimization simulations show that, $\varepsilon_{c,p}$ along zigzag direction is about 0.01, while the buckling does not occur under armchair-along compression even when $\varepsilon \geq 0.05$ (Fig. 3.3). This indicates that, the predictions of Eq. (3.2) based on continuum mechanical theory is no longer available. In this situation, the discontinuous effects should be considered at nanoscale, and the flexural deformation of graphene can no longer be characterized by continuum mechanical models such as plate or membrane [46].

The electron density of graphene in the flexural plane is obtained by DFT simulations (Fig. 3.4a). The electron density of adjacent carbon atoms along the armchair direction forms a cluster. Assuming the angle between adjacent carbon atom columns is $\Delta\varphi$, there is a permissible angle $\Delta\varphi_p$ due to a constrained hinge behavior of the electron cluster (Fig. 3.4a). The flexural angle φ of graphene is the summation of $\Delta\varphi$:

$$\varphi = N \times \Delta\varphi = l/l_{col} \times \varphi, \tag{3.3}$$

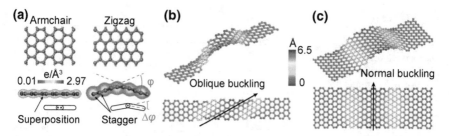

Fig. 3.4 a The electron density of graphene in the flexural plane. Under armchair-along compression, **b** oblique and **c** normal buckling occurs at $\alpha = 0.25$ and $\alpha = 0.5$, respectively

where N is the total number of carbon atom columns, l is the length of graphene sheet, and l_{col} is the length of carbon atom columns. l_{col} along armchair and zigzag direction is 2.13 Å and 1.23 Å, respectively. Due to the hinge effect, $\Delta\varphi$ in armchair direction should be less than the permissible angel $\Delta\varphi_p$:

$$\Delta\varphi = \varphi \frac{l_{col}}{l} < \Delta\varphi_p. \qquad (3.4)$$

According to Eq. 3.4, only when l is large enough, buckling can occur without the hinge effect, which is in agreement with the zero deflection under armchair-along compression. Compared with armchair-along compression, the electron density of adjacent carbon atoms in the flexural plane along the zigzag direction is well staggered (Fig. 3.4a), which does not result hinge effect. Thus, $\Delta\varphi_p$ along zigzag direction is large enough for graphene to buckle at nanoscale. This is consistent with the zigzag-along wrinkles (region A) observed in Fig. 3.1b.

According to Eq. (3.4), $\Delta\varphi$ decreases with the increase of l, which leads to the decrease of the discontinuous effects. MD simulation results suggest that the discontinuous effects are negligible when $l > l_c = 2$ nm (l_c is the critical length). Therefore, in the following discussion of size and chirality effects, l is set to 4 nm to avoid the discontinuous effects. In addition, according to energy minimization simulations, the inhomogeneous stresses due to edge effect [47, 48] result in oblique buckling along armchair direction when aspect ratio $\alpha < 0.5$ (Fig. 3.4b), which can also be considered as a discontinuous effect. As α increases, this edge effect becomes negligible. However, the oblique buckling of graphene sheet does not occur under zigzag-along compression when $\alpha < 0.5$. Energy minimization results show that, at same α ($\alpha < 0.5$), the areal density of strain energy for oblique buckling (armchair-along) and normal buckling (zigzag-along) is 0.0418 eV nm^{-2} and 0.0345 eV nm^{-2}, respectively. Due to these discontinuous effects, when $l < 2.0$ nm or $\alpha < 0.5$, the preferred direction of buckling is zigzag-along.

3.5 The Buckling of Graphene Without Discontinuous Effects

In this section, the buckling of graphene at $\alpha > 0.5$ and $l > 2$ nm is discussed using continuum mechanical models. As analyzed in previous section, discontinuous effects can be neglected in this situation. In addition, the normal buckling at first order is a main concern of buckling problems for the design of graphene-based nanodevices [12, 26]. The maximum deflection $d_{\max}$ of graphene sheet (Fig. 3.2) is important for characterizing the difficulty of buckling, which also plays a dominant role in the design of nanodevices. Based on the plate models in continuum mechanics, the $d_{\max}$ is expressed as [49, 50]:

$$d_{\max} = \frac{2t^e}{\pi c}\sqrt{\varepsilon - \frac{l^2}{\pi^2 wt^e}\frac{D}{C}} = \frac{2t^e}{\pi c}\sqrt{\varepsilon - \frac{l}{\pi^2 \alpha t^e}\frac{D}{C}}, \quad (3.5)$$

where c is a constant, and $t^e = 0.34$ nm is the equivalent thickness of the graphene monolayer. For the first order buckling, $d_{\max}$ is the central deflection of graphene sheet, which characterizes the buckling state. Therefore, at a same l, $d_{\max}$ is jointly determined by ε and α.

Comparing the α–ε diagrams of $d_{\max}$ under armchair- and zigzag-along compression (Fig. 3.5), there exist three stages at different α: (I) $0.5 < \alpha < 1$, (II) $1 < \alpha < 3$, and (III) $\alpha > 3$. At stage I, the maximum deflection under zigzag-along buckling $d_{\max}^{\text{zig}}$ is higher than that under armchair-along buckling $d_{\max}^{\text{arm}}$ at an identical (α, ε): the preferred buckling direction of graphene is zigzag-along, which is consistent with the zigzag-along wrinkles in region A shown in Fig. 3.1b. At stage II, $d_{\max}^{\text{arm}}$ becomes larger than $d_{\max}^{\text{zig}}$ at an identical (α, ε): the preferred growth direction of wrinkles is armchair-along, which is consistent with the armchair-along wrinkles in region B shown in Fig. 3.1b. At stage III, the difference between $d_{\max}^{\text{zig}}$ and $d_{\max}^{\text{arm}}$ is negligible, which is in agreement with the isotropic buckling in large graphene sheet [26].

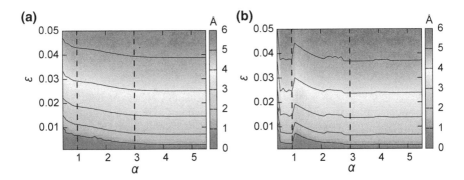

Fig. 3.5 The α–ε diagram of $d_{\max}$ under **a** armchair-along and **b** zigzag-along compression

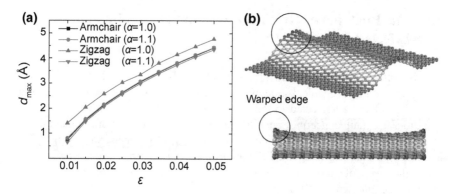

Fig. 3.6 **a** The ε–d_{max} curves at different α. The d_{max}^{zig} at $\alpha = 1.0$ is remarkably larger than that of other three cases. **b** Edges of graphene sheet are warped at $\alpha = 1.0$ under zigzag-along compression

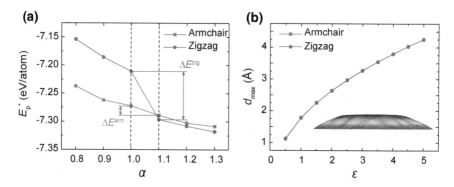

Fig. 3.7 **a** The curves of E_p^* versus α under armchair- and zigzag-along compression. **b** The curves of d_{max} versus ε under armchair- and zigzag-along compression at same α ($\alpha > 3.0$). Inset: The final static buckling pattern of the pre-strained circular graphene monolayer

In addition, there is an unsmooth step of d_{max}^{zig} in the contour lines when α varies from 1.0 to 1.1 (Fig. 3.5b). According to the results of energy minimization simulations, the d_{max}^{zig} at $\alpha = 1.0$ is remarkably larger than that of other three cases (Fig. 3.6a). Actually, for the buckling under zigzag-along compression, edges of graphene sheet are warped at $\alpha = 1.0$ (Fig. 3.6b), which leads to a increase of buckling area. In addition, the contour line of d_{max} (also the axle wires for graphene sheet) is perpendicular to compression direction, which is armchair-along. As analyzed above, discontinuous effects occur along the armchair-along axle wire, which leads to a conjoint increase of deflection. Therefore, the unsmooth step of d_{max}^{zig} is resulted by the warped edges. The effect of warped edges results in the zigzag-along wrinkles at stage I, while can be neglected at $\alpha \geq 1.1$.

The buckling of graphene can also be characterized by potential energy analysis [51]. The initialization of buckling occurs in carbon atoms with high potential energy [52]. At stage I ($0.5 < \alpha \leq 1.0$), the potential energy per atom E_p^* under zigzag-along compression is remarkably larger than that under armchair-along compression due to the effect of warped edges (Fig. 3.7a). At stage II ($1.0 < \alpha < 3.0$), the effect of warped edges decreases, and E_p^* under zigzag-along compression becomes less than that under armchair-along compression due to bending stiffness [43]. In this situation, the preferred growth direction becomes armchair-along, which is consistent with the MD results. At stage III ($\alpha > 3.0$), the ε–d_{max} curves under armchair- and zigzag-along compression are highly coincident (Fig. 3.7b), which indicates an isotropic buckling behavior of graphene. This is consistent with previous research [26, 44]. The final static buckling patterns of the pre-strained circular graphene monolayer with diameter of 60 nm are isotropic (inset in Fig. 3.7b). Therefore, when $\alpha > 3.0$, continuum mechanical models [46] are applicable to characterize the buckling properties of graphene monolayer without discontinuous and chiral effects.

3.6 Chapter Summary

In this chapter, the buckling of graphene monolayer, which is also classified as the static and dynamic out-of-plane mechanical behavior induced by in-plane compression in Chap. 1, is studied. The mechanism of discontinuous effects is revealed, and the applicability of continuum mechanical models is discussed, which is important for the design of graphene and its composites with buckling. The main conclusions can be summarized based on the preferred buckling directions:

- $l < 2$ nm: zigzag-along due to discontinuous effects.
- $l \geq 2$ nm:
  - $\alpha < 0.5$: zigzag-along due to the effects of oblique buckling.
  - $0.5 \leq \alpha \leq 1.0$: zigzag-along due to the effect of warped edges.
  - $1.0 < \alpha \leq 3.0$: armchair-along due to bending stiffness.
  - $\alpha > 3.0$: isotropic.

References

1. Geim AK (2009) Science 324(5934):1530
2. Schedin F, Geim A, Morozov S, Hill E, Blake P, Katsnelson M, Novoselov K (2007) Nat Mater 6(9):652
3. Wang Y, Yang R, Shi Z, Zhang L, Shi D, Wang E, Zhang G (2011) ACS Nano 5(5):3645
4. Heersche HB, Jarillo-Herrero P, Oostinga JB, Vandersypen LM, Morpurgo AF (2007) Nature 446(7131):56
5. Trbovic J, Minder N, Freitag F, Schönenberger C (2010) Nanotechnology 21(27):274005
6. Lu X, Dou H, Gao B, Yuan C, Yang S, Hao L, Shen L, Zhang X (2011) Electrochim Acta 56(14):5115
7. Bunch JS, Verbridge SS, Alden JS, Van Der Zande AM, Parpia JM, Craighead HG, McEuen PL (2008) Nano Lett 8(8):2458
8. Taziev R, Prinz VY (2011) Nanotechnology 22(30):305705

9. Guinea F, Horovitz B, Le Doussal P (2008) Phys Rev B 77(20):205421
10. Guinea F, Horovitz B, Le Doussal P (2009) Solid State Commun 149(27–28):1140
11. De Parga AV, Calleja F, Borca B, Passeggi M Jr, Hinarejos J, Guinea F, Miranda R (2008) Phys Rev Lett 100(5):056807
12. Thompson-Flagg RC, Moura MJ, Marder M (2009) Eur Lett 85(4):46002
13. Huang M, Yan H, Chen C, Song D, Heinz TF, Hone J (2009) Proc Natl Acad Sci 106(18):7304
14. Proctor JE, Gregoryanz E, Novoselov KS, Lotya M, Coleman JN, Halsall MP (2009) Phys Rev B 80(7):073408
15. Xu Y, Gao H, Chen H, Yuan Y, Zhu K, Chen H, Jin Z, Yu B (2012) Appl Phys Lett 100(5):052111
16. Zhu W, Low T, Perebeinos V, Bol AA, Zhu Y, Yan H, Tersoff J, Avouris P (2012) Nano Lett 12(7):3431
17. Morozov S, Novoselov K, Katsnelson M, Schedin F, Elias D, Jaszczak JA, Geim A (2008) Phys Rev Lett 100(1):016602
18. Ramanathan T, Abdala A, Stankovich S, Dikin D, Herrera-Alonso M, Piner R, Adamson D, Schniepp H, Chen X, Ruoff R et al (2008) Nat Nanotechnol 3(6):327
19. Zhang Z, Duan W, Wang C (2012) Nanoscale 4(16):5077
20. Stankovich S, Dikin DA, Dommett GH, Kohlhaas KM, Zimney EJ, Stach EA, Piner RD, Nguyen ST, Ruoff RS (2006) Nature 442(7100):282
21. Rafiee M, Rafiee J, Yu ZZ, Koratkar N (2009) Appl Phys Lett 95(22):223103
22. Paronyan TM, Pigos EM, Chen G, Harutyunyan AR (2011) ACS Nano 5(12):9619
23. Gao L, Guest JR, Guisinger NP (2010) Nano Lett 10(9):3512
24. Wintterlin J, Bocquet ML (2009) Surf Sci 603(10–12):1841
25. Mao Y, Wang WL, Wei D, Kaxiras E, Sodroski JG (2011) ACS Nano 5(2):1395
26. Frank O, Tsoukleri G, Parthenios J, Papagelis K, Riaz I, Jalil R, Novoselov KS, Galiotis C (2010) ACS Nano 4(6):3131
27. Pradhan S (2009) Phys Lett A 373(45):4182
28. Sakhaee-Pour A (2009) Comput Mater Sci 45(2):266
29. Pradhan S, Murmu T (2009) Comput Mater Sci 47(1):268
30. Zhao H, Min K, Aluru N (2009) Nano Lett 9(8):3012
31. Ma T, Li B, Chang T (2011) Appl Phys Lett 99(20):201901
32. Kim DH, Rogers JA (2009) ACS Nano 3(3):498
33. Osvath Z, Lefloch F, Bouchiat V, Chapelier C (2013) Nanoscale 5(22):10996
34. Dragoman D, Dragoman M (2008) Appl Phys Lett 93(10):103105
35. Smolyanitsky A, Tewary VK (2013) Nanotechnology 24(5):055701
36. Plimpton S (1995) J Comput Phys 117(1):1
37. Liew K, Wei J, He X (2007) Phys Rev B 75(19):195435
38. Stuart SJ, Tutein AB, Harrison JA (2000) J Chem Phys 112(14):6472
39. Li J (2003) Model Simul Mater Sci Eng 11(2):173
40. Vanderbilt D (1990) Phys Rev B 41(11):7892
41. Chen Y, Lu J, Gao Z (2007) J Phys Chem C 111(4):1625
42. Wang H, Wang Q, Cheng Y, Li K, Yao Y, Zhang Q, Dong C, Wang P, Schwingenschlogl U, Yang W et al (2011) Nano Lett 12(1):141
43. Giannopoulos GI (2012) Comput Mater Sci 53(1):388
44. Wang H, Upmanyu M (2012) Nanoscale 4(12):3620
45. Timoshenko SP, Gere JM (2009) Theory of elastic stability. Courier Corporation
46. Yue K, Gao W, Huang R, Liechti KM (2012) J Appl Phys 112(8):083512
47. Guo Y, Guo W (2013) Nanoscale 5(1):318
48. Shenoy V, Reddy C, Ramasubramaniam A, Zhang Y (2008) Phys Rev Lett 101(24):245501
49. Rhodes J (2003) Thin-Walled Struct 41(2–3):207
50. Rees DW (2009) Mechanics of optimal structural design: minimum weight structures. Wiley
51. Wang C, Liu Y, Lan L, Tan H (2013) Nanoscale 5(10):4454
52. Runte S, Lazić P, Vo-Van C, Coraux J, Zegenhagen J, Busse C (2014) Phys Rev B 89(15):155427

Chapter 4
Dynamic Ripples in Graphene Monolayer

The dynamic ripples in graphene monolayer are one of the important out-of-plane mechanical behaviors of graphene, which are generally motivated by out-of-plane loading. Their deformation mechanism and potential applications are discussed in this chapter.

4.1 Introduction

Dynamic ripples in graphene [1] are important for potential applications such as resonators [2] and sensors [3] in nanoelectromechanical systems (NEMS) [4], whose fundamental behavior is the transverse wave [5]. At low frequency, the propagation of transverse waves in graphene monolayer is isotropic, which can be characterized by both continuum mechanical models [6] and classical atomistic simulations [7]. Theoretically, the frequency of transverse waves in graphene monolayer can reach terahertz and beyond [8], which brings bright prospects for its applications as terahertz electronic nanodevices [9] and nanoresonators [10]. Compared with the microwaves, terahertz waves in graphene can carry more information and with higher resolution [11]. Although the high frequency vibration of graphene has attracted considerable attention in recent years [12], the continuum equivalent plate models are generally used to study this problem, regardless of possible discontinuous effects [13]. In addition, the effect of chirality can affect the mechanical properties of graphene such as Young's modulus [14] and bending stiffness [15], which is crucial in the design of graphene-based nanodevices [16]. However, neither discontinuous effects nor chiral difference has been considered in the literatures related to the dynamic ripples in graphene monolayer.

Dynamic ripples are not only a fundamental mechanical problem for graphene, but also have potential applications: the ripples can change other physical properties of graphene [17, 18]. The physical manipulations (e.g., opening a band gap in the electronic structure [19], changing the electronic wavefunction [20], and allowing for the observation of excited electron states [21]) make graphene promising in applications as electrochemical capacitors [22, 23], sensors [3], spintronic devices [24], photodetectors [25], and advanced secondary batteries [26]. The theoretical high frequency response also makes graphene an ideal candidate in high frequency nanodevices [27]. With the development of high frequency resonators [10, 28], dynamic ripples in graphene [29] have been applied in numerous dynamic electromechanical systems [30, 31]. The ripples (also called ruga [32]) in graphene can be generated by nanolithography [33], thermal stress [34], nanoparticles [35], and chemical functionalization [36]. The size and quality factor (Q-factor) [37] of graphene ripples [38, 39], are difficult to control during the spontaneous and self-organized formation in these methods above [40]. Prestrain and prestress in graphene monolayer are considered as ideal approaches to control the patterns of ripples [41, 42]. However, effective prestrain or prestress methods for designing dynamic ripple patterns in graphene monolayer have not been reported in the literatures [4, 43].

Experimental results show that, the Q-factor of dynamic ripples in multilayer graphene sheet with a thickness of 20 nm is fine at 50 K [2]. However, the low temperature is impractical for NEMS, and the high thickness of 20 nm decreases the effective frequency of nanodevices [8]. In addition, the size of ripples in multilayer graphene is generally over 10 nm, which is not compatible with the superfine size requirement in nanodevices [32, 40, 44, 45]. The ripples in graphene monolayer can be manipulated by functionalization [36], vacancy clusters [46], electric field [47], defects [48], and grain boundaries [49–51]. However, the distribution of the functional groups, vacancies and grain boundaries in the graphene monolayer are difficult to control during manipulations. Controllable ripples induced by local strain in graphene monolayer have been reported [15, 29, 34, 52, 53], which provides fundamental nature of out-of-plane mechanical behavior of graphene. Therefore, it is possible to obtain well-organized ripples in graphene monolayer by applying stress or strain [32, 54].

In this chapter, the dynamic ripples in graphene monolayer are studied by molecular dynamics (MD) simulations, first-principle calculations, and continuum mechanical models. The effects of the vibrational frequency and chirality on the propagation of transverse waves in graphene monolayer are studied. The phase velocity of transverse waves in graphene increases with vibrational frequency, while there exists a maximum permissible frequency. Transverse waves cannot propagate when the vibrational frequency is over this permissible frequency. In addition, the permissible frequency is chirality-dependent due to discontinuous effects. Based on this understanding, the dynamic ripples induced by transverse waves in graphene monolayer are studied to obtain high-quality and well-organized patterns at room temperature. The dissipation due to edge reflection decreases the quality of ripples, and two methods to generate high-quality ripples are proposed. These high-quality ripples can be accurately controlled by varying the vibrational frequency and interference angle to tailor the electronic properties of graphene, which is important for the design of graphene and its composites in electromechanical systems.

4.2 Transverse Waves in Graphene Monolayer

The mechanical properties of transverse waves in graphene monolayer are discussed in this section. MD simulations are performed by using open source codes LAMMPS [55], and the interactions between carbon atoms are described by the adaptive intermolecular reactive empirical bond order (AIREBO) potential [56]. Before vibrational simulations, all configurations are relaxed at 300 K for 10 ps with an NVT ensemble. The vibration simulations are performed with an NVE ensemble. In order to avoid fracture of graphene, the amplitude of vibration source A_0 is set to 0.01 Å.

4.2.1 Propagation of Transverse Wave in a Square Graphene Sheet

A sinusoidal excitation with vibrational frequency f is applied to the carbon atoms in the central domain of a square graphene sheet with dimensions of $100 \times 100 \, \text{nm}^2$ within a radius of 1 nm. The transverse waves can propagate in-plane freely.

As f increases, the patterns of transverse waves in graphene sheet become ring-like ($f = 0.5 \, \text{THz}$), hexagonal ($f = 6.0 \, \text{THz}$), six-petal-like ($f = 10.5 \, \text{THz}$), and noise ($f = 20.0 \, \text{THz}$), successively (Fig. 4.1). The ring-like pattern indicates that the transverse wave propagation is isotropic, while the hexagonal and six-petal-like patterns demonstrate that there exists chiral difference of propagation. In addition, at a same time step of $t = 6 \, \text{ps}$, the size of hexagonal pattern (Fig. 4.1b) is remarkably larger than that of ring-like pattern (Fig. 4.1a), which suggests that the phase velocity of transverse waves increases with f. This is consistent with the theoretical analysis conducted by Narendar et al. [6]. The hexagonal pattern (Fig. 4.1b) also suggests that the phase velocity is chirality-dependent. As f increases from 6.0 to 20.0 THz, the proportion of noise in ripple patterns increases, which indicates that there exists an upper limitation of f for transverse waves.

According to the MD results, two basic mechanical behaviors of transverse waves in graphene monolayer can be summarized as below:

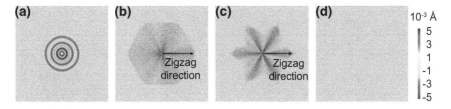

Fig. 4.1 The patterns of transverse waves in the square graphene sheet ($100 \times 100 \, \text{nm}^2$) at $t = 6 \, \text{ps}$ under vibrational frequency of **a** 0.5 THz, **b** 6.0 THz, **c** 10.5 THz, and **d** 20.0 THz. Color bar: out-of-plane displacement

- The phase velocity of transverse waves is chirality-dependent at high frequency.
- There exists an upper limitation of frequency for transverse waves propagation, and the limitation is also chirality-dependent.

4.2.2 The Effects of Chirality and Vibrational Frequency on Transverse Waves Propagation

To study the effects of chirality and vibrational frequency, a one-dimensional (1D) model for transverse waves in graphene monolayer is studied (Fig. 4.2). Periodic boundary conditions are applied along y–axis, and the transverse waves propagate along x–axis. The amplitude of the vibration source is set to 0.01 Å. In this subsection, only the propagation of transverse waves before reflection is discussed.

According to the plate model in continuum mechanics, the 1D propagation of transverse waves in graphene sheet can be characterized as:

$$D\nabla^4 w + \rho h \frac{\partial^2 w}{\partial t^2} = 0, \qquad (4.1)$$

where w is the out-of-plane displacement along z–aixs, D is the bending stiffness, t is the time, ρ and h are the density and thickness of graphene, respectively. For the one-dimensional model, the derivative terms along y–axis are 0. Therefore, the phase velocity of transverse waves c is expressed as [57]:

$$c = \frac{2\pi}{\lambda}\sqrt{\frac{D}{\rho h}}, \qquad (4.2)$$

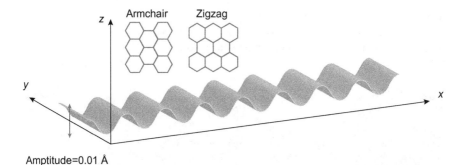

Fig. 4.2 The schematic of 1D model for transverse waves in graphene monolayer. Two different chiralities along x–axis, armchair and zigzag, are simulated

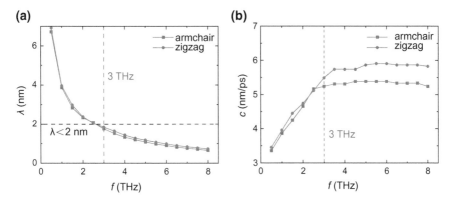

Fig. 4.3 The **a** f–λ and **b** f–c curves

where λ is the wavelength of transverse waves. Substitution of f into λ, Eq. (4.2) can be expressed as:

$$c = f\lambda = \frac{2\pi}{\lambda}\sqrt{\frac{D}{\rho h}} = \omega^{\frac{1}{2}}\left(\frac{D}{\rho h}\right)^{\frac{1}{4}}, \quad (4.3)$$

where $\omega = 2\pi f$ is the angular frequency of transverse waves. In Eq. (4.3), the term ρh can be considered a constant, and the chiral difference of c at same f is generated from the term D. The elastic bending analysis conducted by Giannopoulos et al. [58] shows that, there exits a chiral difference of D at $\lambda < 2$ nm. The f–λ curves along armchair and zigzag directions are approximately coincident (Fig. 4.3a), and f is over 3.0 THz when $\lambda < 2.0$ nm. In this situation, the bending stiffness along zigzag direction is higher than that along armchair due to curvature [15]. Therefore, according to Eq. (4.3), the phase velocity c along zigzag is higher than that along armchair when $f > 3.0$ THz, which is in good agreement with MD results (Fig. 4.3b). In addition, the phase velocity will be approximately a constant at high frequency, which is consistent with the results of Narendar et al. [6].

4.2.3 Discontinuous Effects on the Permissible Frequency

The theoretical frequency response of graphene monolayer is beyond terahertz [59]. However, the transverse waves will be significantly distorted to noise when f is over a critical value, which indicates an upper limitation or permissible frequency f_p. In addition, the six-petal-like pattern of ripples (Fig. 4.1c) suggests that the upper limitation f_p is chirality-dependent. Actually, the patterns of ripples can be characterized by the resonant amplitude of wavefront A_r. In general, the transverse waves can be defined as distorted when their normalized resonant amplitude $A_r^* = A_r/A_0$ (A_0 is the amplitude of vibrational source) is less than 5% [60].

Four stages can be classified according to the f–A_r^* curves (Fig. 4.4a):

- Stage I ($f < 3.0$ THz): the f–A_r^* curves along armchair and zigzag direction are almost coincident, which indicates that the propagation of transverse waves is isotropic in graphene monolayer.
- Stage II (3.0 THz $< f < 10.0$ THz): the resonant amplitude of transverse waves along armchair and zigzag directions is comparable with A_0, but the phase velocity becomes chirality-dependent due to the effect of bending stiffness as analyzed before.
- Stage III (10.0 THz $< f < 16.0$ THz): the waveform declines to noise along armchair direction, while it is still normal along zigzag direction.
- Stage IV ($f > 16.0$ THz): the waveforms in both armchair and zigzag directions decline to noise.

Actually, the four patterns of ripples in Fig. 4.1 can be completely corresponding to the four stages above. In addition, the permissible frequency f_p along armchair and zigzag direction is about 10.0 THz and 16.0 THz, respectively. When $f > f_p$, the waveform declines to noise. For example, at $f = 10.5$ THz, the waveform of transverse waves along armchair direction declines to noise, while that along zigzag direction is still normal (Fig. 4.4b).

When λ is close to the C–C bond length, the discontinuous effects should be considered, and the out-of-plane vibration of graphene should be considered as the vibration of discrete nodes. For these discrete nodes, if there are less than 3 nodes in a single wavelength, there must exist a time that all nodes have a same velocity component. In this situation, the vibration becomes in rigid-body mode with a frequency of 0 Hz, which leads to the disappearance of transverse waves. Therefore, to ensure the propagation of transverse waves in graphene monolayer, the minimum permissible wavelength should include at least 3 nodes. A single node is consisting of 2 rows of carbon atoms along armchair direction (Fig. 4.5a), while that is consisting of 1

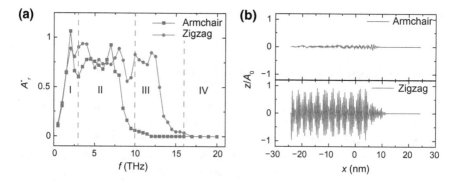

Fig. 4.4 The f–A_r^* curves at different chiralities. **b** Distorted waveform occurs along armchair direction but not along zigzag direction at $f = 10.5$ THz

4.2 Transverse Waves in Graphene Monolayer

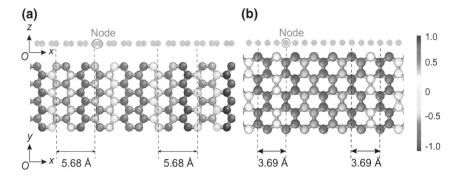

Fig. 4.5 The discrete vibrational models for graphene monolayer along **a** armchair and **b** zigzag direction. Color bar: normalized out-of-plane displacement z/A_0. The permissible wavelengths along armchair and zigzag direction are 5.68 Å and 3.69 Å, respectively

row of carbon atoms along zigzag direction (Fig. 4.5b). This is consistent with the electron density analysis in Chap. 3. Therefore, the permissible wavelengths along armchair λ_p^{arm} and zigzag λ_p^{zig} directions are 5.68 Å and 3.69 Å (Fig. 4.5), respectively. According to Eq. (4.3), the permissible frequency f_p is expressed as:

$$f_p = c/\lambda_p. \tag{4.4}$$

According to Eq. (4.4), the theoretical permissible frequency $f_p^{arm} = 9.4$ THz, and $f_p^{zig} = 15.9$ THz, respectively. This is in good agreement with the MD results. Therefore, the upper limitation of frequency of transverse waves in graphene monolayer is resulted by the discontinuous effect.

4.3 Controllable Dynamic Ripples in Graphene Monolayer

Through analyzing the mechanical properties of transverse waves in graphene, the underlying mechanisms of anisotropic propagation and permissible frequency have been revealed. Based on these mechanisms, the controllable dynamic ripples generated by the interference of transverse waves in graphene monolayer is studied in this section.

4.3.1 Simulation Details

MD simulations are performed using LAMMPS [55]. The interactions between carbon atoms are described by AIREBO potential [56]. All initial configurations are relaxed for 100 ps at 300 K with NVT ensemble, and interference simulations are performed with NVT ensemble. In reality, the graphene sheet is generally supported or fixed by other devices. Therefore, a 3D harmonic position restraint at the boundary is set by applying a spring force independently to each boundary atom to tether it to its initial position. Assuming the edge atoms are connected with SiO_x substrates, the spring constant is $0.5\,eV/Å^2$ (adhension energy $\sim 10\,J/m^2$) [61], and a Langevin thermostat at 300 K is applied to these atoms. For all atomistic configurations, the width of edge is set to 1.0 nm. Point vibrational sources are simulated by applying sinusoidal excitation to atoms in a central domain within a radius of 1.0 nm. Although the free-standing portion of graphene may interact with its surroundings (e.g., gas atoms, light, electronic and magnetic field) in NEMS, experimental and theoretical results [50, 62] show that these interactions can be neglected compared with the mechanical excitation. Therefore, the graphene sheet is placed in a vacuum. The time step is set to 1.0 fs. Visualization is performed by using AtomEye [63].

The electronic properties of ripples are studied in this section, and the related DFT simulations are performed using an open source code QUANTUM-ESPRESSO [64]. PBE exchange-correlation, Vanderbilt ultrasoft pseudopotentials, and a kinetic energy cutoff energy of 20 Ry are used in these calculations [65]. Each supercell should be compatible with graphene lattice, and the vacuum space along out-of-plane direction is set to 4.0 nm to avoid self-interactions.

4.3.2 Patterns and Quality of Dynamic Ripples Generated by the Interference of Transverse Waves

In NEMS, the harmonic out-of-plane excitation is realized by traveling nanoindentations and beams of coherent acoustic phonons stimulated by interdigitated transducers [66]. Actually, these nanoindentations and beams can be considered as point vibrational sources (PVSes) and line vibrational sources (LVSes), respectively. Through the interferences of transverse waves in graphene monolayer, three representative patterns of dynamic ripples can be generated (Fig. 4.6). For the interference between two LVSes, the patterns of dynamic ripples are generally rectangle or parallelogram (Fig. 4.6a), which are determined by the interference angle and frequency of LVS. When the transverse waves from a LVS encounter a fixed boundary, they are reflected and transform to another series of waves with same frequency. Therefore, similar to the double-slit experiment, axisymmetric patterns of dynamic ripples can be obtained through the self-interference of transverse waves from a single LVS (Fig. 4.6b). Radial patterns are generated by the interference of two PVSes (Fig. 4.6c). For the transverse waves from PVSes, it is difficult to obtain the latticed patterns of dynamic ripples. In addition, compared with LVS at same amplitude, the out-of-plane displacement of dynamic ripples generated by PVS is much smaller (Fig. 4.6c). Actually, the dynamic ripples generated by LVSes are mainly used to improve the electronic properties such as band gap engineering [34], and the spacial quality factor (Q_s) is the main concern; on the other hand, the dynamic ripples

4.3 Controllable Dynamic Ripples in Graphene Monolayer

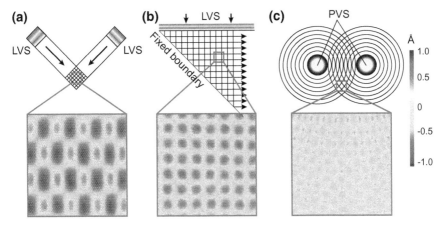

Fig. 4.6 Three representative patterns of dynamic ripples generated by interferences of **a** two LVSes, **b** self-interference of a single LVS, and **c** two PVSes. Color bar: out-of-plane displacement

generated by PVSes are mainly used in sensor devices, which pay more attention to the temporal quality factor (Q_t).

Stotz et al. have realized the coherent spin transport through dynamic quantum dots generated by interference of transverse waves in undoped GaAs (001) quantum wells [62], and the average size of these ripples is over 1 μm. Similar method is applied to graphene monolayer by MD simulations to obtain dynamic ripples at nanoscale. However, irregular and unstable ripples form after the reflection of transverse waves at edges (Fig. 4.7). In this situation, these ripples with low Q_s and Q_t are not compatible with high frequency devices in NEMS [67].

Theoretically, the unstable and low-quality ripples are resulted by the interference of two transverse waves with different frequency f, which occur after the reflection at edges due to dispersion [7]. If the propagation direction of these dispersed waves is directed away from the initial waves, this effect can be weakened. Similar to the propagation of light in a multi-mode optical fiber, a parallelogram graphene sheet is designed to make the transverse waves propagate along a one-way path (red arrows in Fig. 4.8a). MD results show that latticed dynamic ripples can be obtained in the middle region of graphene sheet after 100 ps, while the dispersed ripples are generated at the right end (Fig. 4.8a). The the envelope line of waveform $w(x)$ on the central axis of the graphene sheet can be fitted by an exponential function (Fig. 4.8b):

$$f(x) = e^{-\xi kx} = e^{-2\pi \xi x/\lambda}, \quad (4.5)$$

where λ is the lattice parameter of the latticed ripples, and ξ is a dimensionless parameter. The spatial Q-factor Q_s is given as [50, 68]:

$$Q_s = 0.5/\xi. \quad (4.6)$$

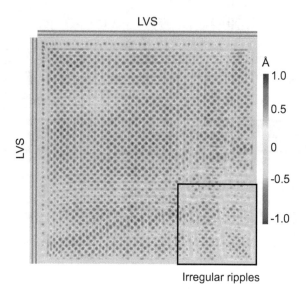

Fig. 4.7 Irregular and unstable ripples form after the reflection of transverse waves at edges. Color bar: out-of-plane displacement

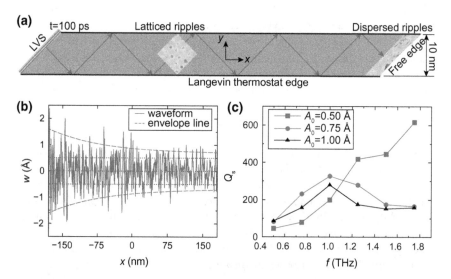

Fig. 4.8 **a** Parallelogram graphene sheet with latticed dynamic ripples at $t = 100$ ps; **b** An example of dynamic ripples at $f = 1.0$ THz and $A_0 = 1.0$ Å: the envelope line of the waveform can be fitted by the exponential curve $e^{-\xi k x}$; **c** Q_s–f curves at different A_0

4.3 Controllable Dynamic Ripples in Graphene Monolayer

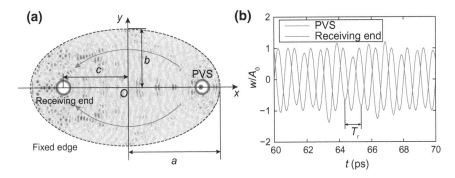

Fig. 4.9 **a** Dynamic ripples in an elliptical graphene sheet ($x^2/a^2 + y^2/b^2 = 1$) with a PVS at the right focal point $(c, 0)$, and the receiving end is the left focal point $(-c, 0)$; **b** The vibrational curves of normalized out-of-plane deflection w/A_0 for the PVS and receiving end

The Q_s increases with frequency f when the amplitude of LVS $A_0 = 0.5$ Å. However, when $A_0 = 0.75$ Å or $A_0 = 1.00$ Å, the Q_s increases first, then decreases, and reaches its maximum at $f = 1.0$ THz, which will be analyzed later. Through this method, the dynamic ripples with $Q_s > 100$ can be obtained at room temperature, and the average size of these ripples is about 1.0 nm, which are desired and practical in NEMS [7].

On the other hand, the temporal Q-factor Q_t of dynamic ripples is important for the point to point synchronizations of devices in NEMS [69], and these ripples can be generated by PVSes. In general, the signal strength of waves will be weakened due to dissipation. If the waves with an identical frequency can reach the receiving end continuously, the dissipation effect may be neglected. Similar to the convergence of light in an ellipse, an elliptical graphene sheet with a PVS at its right focal point is set to obtain dynamic ripples with high Q_t (Fig. 4.9a). In this situation, the waves are reflected and then focus on the left focal point (the receiving end) in the elliptical graphene sheet. The Q_t is expressed as:

$$Q_t = \frac{T_s}{|T_s - T_r|}, \tag{4.7}$$

where T_s and T_r are the vibrational period of PVS and receiving end. The MD results show that T_r and T_s are almost identical, which leads to a high Q_t of 10^5.

4.3.3 The Motion and Electronic Properties of Dynamic Ripples

Assuming the dynamic ripples are generated by LVS in an infinite plane regardless of the reflection at edges, the phase velocity of the ripple c_r is expressed as:

$$c_{\rm r} = c_{\rm r}(f) = \frac{c_{\rm s}(f)}{\sin(\theta/2)}, \quad (4.8)$$

where $c_{\rm s}$ is the phase velocity of transverse waves generated by vibrational source, which is a function of vibrational frequency f as analyzed in Sect. 4.2, and θ is the interference angle. For the latticed dynamic ripples generated by LVSes (Fig. 4.10), their sizes can be characterized by two parameters λ_x and λ_y in xy plane (Fig. 4.10a), which are determined by f and θ:

$$\lambda_x \sin(\theta/2) = \lambda_y \cos(\theta/2) = \lambda(f), \quad (4.9)$$

MD results show that, when $f < 1.0$ THz, the phase velocity of ripples is approximately a constant at different θ (Fig. 4.11a). In this situation, substitution of Eq. (4.8) into Eq. (4.9), λ_y can be expressed as an inverse function:

$$\lambda_y = \frac{\lambda(f)}{\cos(\theta/2)} = \frac{c_{\rm s}}{f \cos(\theta/2)}. \quad (4.10)$$

The MD results are in good agreement with the theoretical solutions (Fig. 4.11b). However, MD results show that, when $f > 1.0$ THz, $c_{\rm r}$ increases with f, and λ_y can

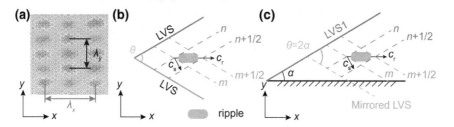

Fig. 4.10 **a** The xy dimensions of a single ripple in graphene monolayer. The motion of the ripple generated by **b** the interference between two LVSes and **c** the self-interference of a single LVS

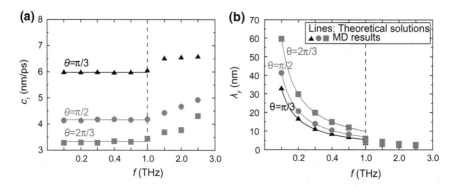

Fig. 4.11 The **a** f–$c_{\rm r}$ and **b** f–λ_y curves of dynamic ripples at $\theta = \pi/3, \pi/2$ and $2\pi/3$

4.3 Controllable Dynamic Ripples in Graphene Monolayer

no longer be fitted by the theoretical solutions. Actually, the dissipation of transverse waves is proportional to $(A/\lambda)^2$ [7], and there exists a significant chiral difference of flexural energy when the flexural length is less than 2 nm [58]. When the $f > 1.0$ THz, the size of dynamic ripples becomes less than 2 nm, and the dissipation effect leads to a decrease of Q_s. In addition, decreasing A leads to a decrease of dissipation effect. This explains the f–Q_s curves shown in Fig. 4.8c.

Although terahertz transverse waves are considered promising in NEMS applications [59], MD results show that the dynamic ripples are not compatible with these nanodevices at $f > 3.0$ THz. When $f > 3.0$ THz, even without dissipation or dispersion effects, the the dynamic ripples are disordered (Fig. 4.12). As mentioned in Sect. 4.2, there exists a significant chiral difference of phase velocity of transverse waves when $f > 3.0$ THz, which results a series of waves with different wavelength. The intrinsic dispersion of waves due to the chiral difference of wavelength deteriorates the distribution of dynamic ripples. Fortunately, even without this effect, the nominal size of dynamic ripples is less than 1.0 nm at $f > 3.0$ THz, which is not compatible with most nanodevices due to strong discontinuous effects [62].

The band gap of graphene monolayer can be opened by the ripples, which is affected by the size of ripples [19]. The nominal radius R of dynamic ripple is defined as:

$$R = \sqrt{\lambda_x \lambda_y / \pi} = \lambda \sqrt{2/(\pi \sin \theta)}. \quad (4.11)$$

The DFT results show that, at $A = 0.5$ Å and $\theta = \pi/2$, as frequency f increases, the nominal radius of dynamic ripple R decreases, and the band gap increases (Fig. 4.13). Actually, the tensile strain in a single ripple is estimated as A/R. The results of Ni et al. show that, a uniaxial tensile strain of 1% can lead to a increase of band gap by 0.3 eV [70], which is consistent with the DFT results in this section. In this situation, the band gap of ripples in graphene monolayer U can be tailored by f and

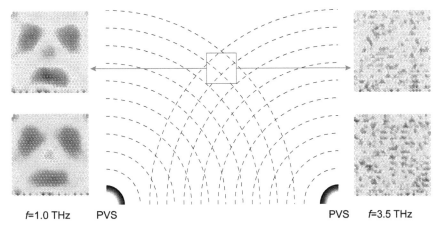

Fig. 4.12 The dynamic ripples in graphene monolayer are ordered at $f = 1.0$ THz, while disordered at $f = 3.5$ THz

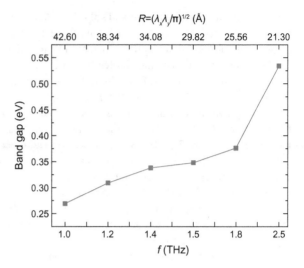

Fig. 4.13 The band gap of a single ripple in graphene monolayer increases with f

θ. Therefore, at an optional amplitude (e.g., $A = 0.5$ Å), desired band gap of ripples can be obtained by varying the vibrational frequency and interference angle, which is important for the applications in NEMS.

4.4 Chapter Summary

In this chapter, the dynamic ripples in graphene monolayer generated by out-of-plane loading is discussed. The basic conclusions are listed below:

- The propagation of transverse waves in graphene monolayer is isotropic when the frequency is less than 3.0 THz, while anisotropic when the frequency is over 3.0 THz due to the chiral difference of bending stiffness.
- The permissible frequencies for transverse waves in graphene monolayer are about 10.0 THz and 16.0 THz along armchair and zigzag direction, respectively. Actually, when the wavelength is close to the lattice of graphene, the discontinuous effect plays a dominant role, and the continuum plate models are no longer applicable.
- High-quality ripples can be obtained by the interference of transverse waves in graphene monolayer at room temperature. The desired electronic properties of dynamic ripples in graphene monolayer can be achieved by tailoring the vibrational frequency and interference angle.
- Considering the effects of anisotropic propagation, dispersion and dissipation, the practicable vibrational frequency for generating dynamic ripples in graphene monolayer in NEMS is about 1.0~3.0 THz.

Up to the present chapter, the three basic out-of-plane mechanical behaviors of graphene raised in Chap. 1 have been studied, and the fundamental mechanisms are revealed. It is noted that, the discontinuous effects exist in the out-of-plane deformation of graphene induced by both in-plane and out-of-plane loading. These effects observed in Chaps. 3 and 4 can be briefly summarized as (which may be universal for all materials):

"When the characteristic dimensions of mechanical behavior are close to the size of basic unit of the material, the materials should be considered discontinuous along the characteristic direction, and the discontinuous effects play a dominant role in related mechanical properties."

These mechanical properties are important for the design of graphene as well as its composites, which will be discussed in the following chapters.

References

1. Geim AK (2009) Science 324(5934):1530
2. Bunch JS, Van Der Zande AM, Verbridge SS, Frank IW, Tanenbaum DM, Parpia JM, Craighead HG, McEuen PL (2007) Science 315(5811):490
3. Robinson JT, Perkins FK, Snow ES, Wei Z, Sheehan PE (2008) Nano Lett 8(10):3137
4. Bunch JS, Verbridge SS, Alden JS, Van Der Zande AM, Parpia JM, Craighead HG, McEuen PL (2008) Nano Lett 8(8):2458
5. Rangel NL, Seminario JM (2008) J Phys Chem A 112(51):13699
6. Narendar S, Gopalakrishnan S (2010) Physica E Low Dimens Syst Nanostruct 43(1):423
7. Kim SY, Park HS (2011) J Appl Phys 110(5):054324
8. Shi JX, Ni QQ, Lei XW, Natsuki T (2011) J Appl Phys 110(8):084321
9. Han MY, Özyilmaz B, Zhang Y, Kim P (2007) Phys Rev Lett 98(20):206805
10. Barton RA, Ilic B, Van Der Zande AM, Whitney WS, McEuen PL, Parpia JM, Craighead HG (2011) Nano Lett 11(3):1232
11. Sensale-Rodriguez B, Yan R, Kelly MM, Fang T, Tahy K, Hwang WS, Jena D, Liu L, Xing HG (2012) Nat Commun 3:780
12. Chowdhury R, Adhikari S, Scarpa F, Friswell M (2011) J Phys D Appl Phys 44(20):205401
13. Narendar S, Mahapatra DR, Gopalakrishnan S (2010) Comput Mater Sci 49(4):734
14. Zhao H, Min K, Aluru N (2009) Nano Lett 9(8):3012
15. Ma T, Li B, Chang T (2011) Appl Phys Lett 99(20):201901
16. Kim SY, Park HS (2009) Nano Lett 9(3):969
17. Miranda R, de Parga ALV (2009) Nat Nanotechnol 4(9):549
18. Lin SY, Chang SL, Shyu FL, Lu JM, Lin MF (2015) Carbon 86:207
19. Kong EH, Joo SH, Park HJ, Song S, Chang YJ, Kim HS, Jang HM (2014) Small 10(18):3678
20. Barnard AS, Snook IK (2012) Nanoscale 4(4):1167
21. Güttinger J, Molitor F, Stampfer C, Schnez S, Jacobsen A, Dröscher S, Ihn T, Ensslin K (2012) Rep Prog Phys 75(12):126502
22. Simon P, Gogotsi Y (2010) Nanoscience and technology: a collection of reviews from nature journals. World Scientific, pp 320–329
23. Hu Y, Zhao Y, Lu G, Chen N, Zhang Z, Li H, Shao H, Qu L (2013) Nanotechnology 24(19):195401
24. Wolf S, Awschalom D, Buhrman R, Daughton J, Von Molnar S, Roukes M, Chtchelkanova AY, Treger D (2001) Science 294(5546):1488
25. Kim CO, Hwang SW, Kim S, Shin DH, Kang SS, Kim JM, Jang CW, Kim JH, Lee KW, Choi SH et al (2014) Sci Rep 4:5603
26. Miller JR, Simon P (2008) Sci Mag 321(5889):651
27. Smolyanitsky A, Tewary VK (2013) Nanotechnology 24(5):055701

28. Huang XMH, Zorman CA, Mehregany M, Roukes ML (2003) Nature 421(6922):496
29. He Y, Li H, Si P, Li Y, Yu H, Zhang X, Ding F, Liew KM, Liu X (2011) Appl Phys Lett 98(6):063101
30. Astley M, Kataoka M, Ford C, Barnes C, Anderson D, Jones G, Farrer I, Ritchie D, Pepper M (2007) Phys Rev Lett 99(15):156802
31. Yoshida K, Xudong Z, Bright AN, Saitoh K, Tanaka N (2013) Nanotechnology 24(6):065705
32. Zhang T, Li X, Gao H (2014) J Mech Phys Solids 67:2
33. Puddy RK, Chua C, Buitelaar M (2013) Appl Phys Lett 103(18):183117
34. Bao W, Miao F, Chen Z, Zhang H, Jang W, Dames C, Lau CN (2009) Nat Nanotechnol 4(9):562
35. Osváth Z, Gergely-Fülöp E, Nagy N, Deák A, Nemes-Incze P, Jin X, Hwang C, Biró LP (2014) Nanoscale 6(11):6030
36. Ding Y, Cheng H, Zhou C, Fan Y, Zhu J, Shao H, Qu L (2012) Nanotechnology 23(25):255605
37. Alivisatos AP (1996) Science 271(5251):933
38. Politano A, Chiarello G (2014) Nanoscale 6(19):10927
39. Kim S, Hee Shin D, Oh Kim C, Seok Kang S, Min Kim J, Choi SH, Jin LH, Cho YH, Won Hwang S, Sone C (2012) Appl Phys Lett 101(16):163103
40. Antonova IV, Nebogatikova NA, Prinz VY (2014) Appl Phys Lett 104(19):193108
41. Bonfanti M, Casolo S, Tantardini GF, Ponti A, Martinazzo R (2011) J Chem Phys 135(16):164701
42. Bonilla L, Carpio A (2012) Phys Rev B 86(19):195402
43. Fricke L, Wulf M, Kaestner B, Kashcheyevs V, Timoshenko J, Nazarov P, Hohls F, Mirovsky P, Mackrodt B, Dolata R et al (2013) Phys Rev Lett 110(12):126803
44. Jin SH, Kim DH, Jun GH, Hong SH, Jeon S (2013) ACS Nano 7(2):1239
45. Liang G, Dupont E, Fathololoumi S, Wasilewski ZR, Ban D, Liang HK, Zhang Y, Yu SF, Li LH, Davies AG et al (2014) Sci Rep 4:7083
46. Singh AK, Penev ES, Yakobson BI (2010) ACS Nano 4(6):3510
47. Osvath Z, Lefloch F, Bouchiat V, Chapelier C (2013) Nanoscale 5(22):10996
48. Zhang T, Gao H (2015) J Appl Mech 82(5):051001
49. Capasso A, Placidi E, Zhan H, Perfetto E, Bell JM, Gu Y, Motta N (2014) Carbon 68:330
50. Qi Z, Park HS (2012) Nanoscale 4(11):3460
51. Yi L, Yin Z, Zhang Y, Chang T (2013) Carbon 51:373
52. Fuhrmann DA, Thon SM, Kim H, Bouwmeester D, Petroff PM, Wixforth A, Krenner HJ (2011) Nat Photonics 5(10):605
53. Wang C, Liu Y, Lan L, Tan H (2013) Nanoscale 5(10):4454
54. Dong Y, He Y, Wang Y, Li H (2014) Carbon 68:742
55. Plimpton S (1995) J Comput Phys 117(1):1
56. Stuart SJ, Tutein AB, Harrison JA (2000) J Chem Phys 112(14):6472
57. Graff KF (2012) Wave motion in elastic solids. Courier Corporation
58. Giannopoulos GI (2012) Comput Mater Sci 53(1):388
59. Lovat G, Burghignoli P, Araneo R (2013) IEEE Trans Electromagn Compat 55(2):328
60. Hunter I (2001) Theory and design of microwave filters, vol 48. Iet
61. Das S, Lahiri D, Agarwal A, Choi W (2014) Nanotechnology 25(4):045707
62. Stotz JA, Hey R, Santos PV, Ploog KH (2005) Nat Mater 4(8):585
63. Li J (2003) Modell Simul Mater Sci Eng 11(2):173
64. Giannozzi P, Baroni S, Bonini N, Calandra M, Car R, Cavazzoni C, Ceresoli D, Chiarotti GL, Cococcioni M, Dabo I et al (2009) J Phys Condens Matter 21(39):395502
65. Cao C, Wu M, Jiang J, Cheng HP (2010) Phys Rev B 81(20):205424
66. Breitwieser R, Hu YC, Chao YC, Li RJ, Tzeng YR, Li LJ, Liou SC, Lin KC, Chen CW, Pai WW (2014) Carbon 77:236
67. Kashcheyevs V, Kaestner B (2010) Phys Rev Lett 104(18):186805
68. Vallabhaneni AK, Rhoads JF, Murthy JY, Ruan X (2011) J Appl Phys 110(3):034312
69. Kataoka M, Fletcher J, See P, Giblin S, Janssen T, Griffiths J, Jones G, Farrer I, Ritchie D (2011) Phys Rev Lett 106(12):126801
70. Ni ZH, Yu T, Lu YH, Wang YY, Feng YP, Shen ZX (2008) ACS Nano 2(11):2301

Chapter 5
Defect-Induced Discontinuous Effects in Graphene Nanoribbon Under Torsion Loading

Defects are ubiquitous in graphene monolayer, which are considered as the foundation of the design of graphene and graphene composites. In addition, defects may induce discontinuous effects for the mechanical behaviors of graphene. In this chapter, the defect technology of graphene is reviewed and discussed. The mechanical behaviors of graphene nanoribbon under torsion loading, which is a representative case including both the static and dynamic out-of-plane deformation, is analyzed to reveal the discontinuous effects induced by defects.

5.1 Introduction

Compared with other common materials, graphene can be designed at atomistic scale based on defect technology. Defects and vacancies are ubiquitous in graphene monolayer [1], which can change its mechanical [2], electrical [3] and magnetic [4, 5] properties. In recent years, desired defects in graphene are mainly obtained by the irradiation of ions [6], electrons [7, 8] and atoms [9], which are the foundation of the design of graphene and graphene composites. The defect controllability of graphene is one of the most concerned issues in this field.

In general, the pattern and formation probability are used to characterize the controllability of inducing defects in graphene. In 2012, Wang et al. proposed a "two-step" experimental method to dope desired atoms into graphene [10]: first generate vacancies by high-energy atom/ion bombardment, and then fill these vacancies with the low-energy desired atoms. Krasheninnikov et al. studied the formation and substitution probability of defects in graphene under atomistic bombardments via classical MD simulations [11–13], and Bubin et al. revealed the atomistic mechanical process via time-dependent density functional theory (TDDFT) simulations [14]. However, the impact site and the physical properties of incident atoms are not discussed in these literatures, and a brief discussion on this point will be conducted in this chapter. Although this issue is not the focus of this thesis, it is still necessary to provide a brief discussion and review, since it is highly related to the controllable grain boundaries in following sections.

In materials science, grain boundaries are defined as the 2D defects in the crystal structure, and tend to decrease the electrical and thermal conductivity of the material [15]. Since graphene is a 2D material, its defects are actually equivalent to grain boundaries. In general, defects are considered a disadvantage for the mechanical properties of materials. However, the results of Ruoff and Wei et al. show that the effects of defects on in-plane tensile strength of graphene is negligible [2, 16]. In addition, as analyzed in previous chapters, defects may increase the discontinuous effects of graphene.

According to the results and analysis in previous chapters, the discontinuous effects on the static and dynamic out-of-plane deformation of graphene are similar. Therefore, graphene nanoribbons (GNRs) under torsion loading are studied in this chapter to reveal the defect-induced discontinuous effects, which is a representative case including both the static and dynamic out-of-plane deformation.

5.2 The Defect Controllability of Graphene Under Atomistic Bombardment

5.2.1 MD Simulation Details

In previous research, classical MD and TDDFT methods have been used to study the atomistic bombardment of graphene [13, 14]. For classical MD simulations, the final patterns of defects can be obtained after long enough relaxation, while the atomistic information such as bonding cannot be obtained due to the inaccuracy of potentials. For TDDFT simulations, the atomistic information can be obtained, while the final pattern of defects cannot be obtained due to a high computation cost. Therefore, in this section the carbon-carbon and carbon-projectile interactions are described by the reactive force field potential (ReaxFF) [17]. The accuracy of ReaxFF is close to that of DFT simulations with much lower computation cost [18, 19]. Bond order and bond energy can be obtained by ReaxFF potential. Three representative atoms, gold (Au) [20, 21], iron (Fe) [22], and oxygen (O) [23], are selected as incident atoms, and their electronegativities increase with the decrease of atomic mass. The MD simulations based on ReaxFF are performed by LAMMPS [24], and the visualizations are performed by AtomEye [25].

An incident atom with initial kinetic energy is placed above the center of the suspended graphene sheet to generate defects (Fig. 5.1). The dimensions of graphene sheet are 10×10 nm^2 consisting of $\sim$3,000 atoms. The initial configuration of graphene sheet is relaxed with NVT ensemble for 10 ps, and the bombardment is simulated with NVE ensemble. After the bombardment, the graphene is relaxed with NVT ensemble for 4 ps to obtain the final pattern of defects. The time step of

5.2 The Defect Controllability of Graphene Under Atomistic Bombardment

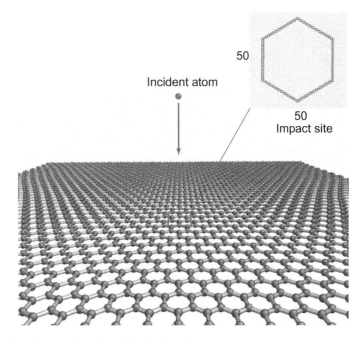

Fig. 5.1 Atomistic configuration of the bombardment process

MD simulations is set to 0.1 fs to obtain the bonding and debonding information. Since the energy for a single carbon atom escaping from sp^2 structure in graphene is about 20 eV, the minimum kinetic energy of incident atoms E_k^i is set to 10 eV. As mentioned in previous section, the impact site has rarely been considered. In this section, a rectangular unit cell consisting of 51×51 grid points, which contains a single graphene hexagon, are used to consider the effect of impact site. A series of bombardment simulations are performed at different grid points.

5.2.2 The Effect of Impact Site

For the Fe-incident bombardments at $E_k^i = 100$ eV, over 90% of the impact sites would not result defects (Fig. 5.2a), which is consistent with the defecting probability less than 10% in previous studies [11–13]. In general, according to the number of escaped carbon atoms N_{esc}, the defects induced by bombardments can be classified as single-vacancy (SV, $N_{esc} = 1$), double-vacancy (DV, $N_{esc} = 2$), treble-vacancy (TV, $N_{esc} = 3$) and multi-vacancy (MV, $N_{esc} > 3$), respectively. Actually, defects are generated only when impact site is in the region where the electron density is over 1.5 e/Å^3 (Fig. 5.2b). High electron density implies a strong interaction between the incident and carbon atoms, and thus the covalent C–C bonds can be destroyed.

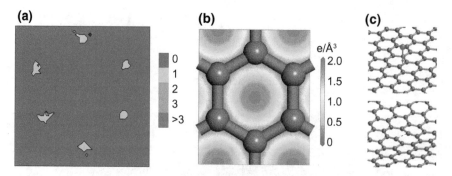

Fig. 5.2 **a** The diagram of N_{esc} at different impact site. **b** The in-plane electron density of graphene in a hexagon. **c** For the head-to-bond bombardment, a metastable bond-vacancy forms without escaped carbon atoms

These impact sites are referring to head-to-head and head-to-bond bombardments. However, in some cases, although a metastable bond-vacancy forms after head-to-bond bombardment, it can be healed after relaxation (Fig. 5.2c).

MD results show that (Fig. 5.2a), as N_{esc} increases, the related area of defect decrease, which is in good agreement with the probability distribution proposed by Krasheninnikov et al. [12, 13]. In addition, there are three stages with the increase of E_k^i:

- $E_k^i < E_{k,p}^i$: defects cannot be generated. Here $E_{k,p}^i$ is the minimum incident energy to generate defects in graphene monolayer.
- $E_{k,p}^i < E_k^i < E_{k,s}^i$: SV can be generated. Here $E_{k,s}^i$ is the maximum incident energy to generate SV. At this stage, the defect probability increases with E_k^i.
- $E_k^i > E_{k,s}^i$: TV or MV are generated randomly.

Similar phenomena are obtained for Au- and O-incident bombardments. Although the defect probability is less than 10%, the sp^2 structure of graphene will not be destroyed in the other 90% bombardments without generating defects. Therefore, multiple bombardments can be performed until the desired defects in graphene are obtained.

5.2.3 The Effects of Physical Properties and Kinetic Energy of Incident Atom

As analyzed in previous subsection, defects are generally generated only under head-to-head and head-to-bond bombardments. In addition to the impact site, the physical properties and kinetic energy of incident atom are also important. Actually, the physical properties of incident atom will highly affect the bonding and debonding with

5.2 The Defect Controllability of Graphene Under Atomistic Bombardment

Table 5.1 Bonding information between incident and carbon atoms

Incident atom	Maximum charge variation (e⁻)	Average bond order
O	−0.3	2.737
Fe	+0.5	0.305
Au	∼0	∼0

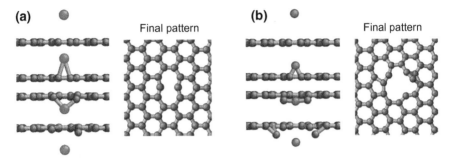

Fig. 5.3 The **a** Fe and **b** O bombardment processes at E_k^i = 100 eV. The defects in graphene generated by O bombardment is random and disordered

carbon atoms. According to ReaxFF calculations, the bonding information between incident and carbon atoms are listed in Table 5.1.

The interatomic forces increase with the bond order, which plays an important role in the final pattern and quality of defects during bombardments. For the Au–C interatomic forces, there is no charge transfer and their bond order is ∼0, which is a Van der Waals force without any chemical bonds. For the O–C interatomic forces, the bond order is much higher than the C–C bond order (∼1.3) with transferring 0.3 electron from C to O, which is a strong chemical bond. Compared with the two interactions, the Fe–C interatomic force is an interaction between Van der Waals force and chemical bond.

MD results show that, these interatomic forces highly affect the quality of defects. For the head-to-bond bombardments, only Fe and O atoms bond with C atoms (Fig. 5.3), and there does not exist chemical bonds between Au and C atoms. Since the Fe–C bond order is less than that of C–C bonds, the carbon atoms will not be dragged by Fe atom, which results a metastable bond-vacancy (Fig. 5.3a). However, the carbon atoms will be strongly dragged by O atom due to a high bond order, which results random and disordered defects in graphene monolayer (Fig. 5.3b). For generating defect in graphene, the controllability of the latter is obviously very low due to the random dragging.

Since the most of forces are concentrated on the impacted carbon atom, atom-drag phenomena observed in head-to-bond bombardments does not occurs for head-to-head bombardment. However, in this situation, the bonds between incident and carbon atoms are still important: MD results show that a one-step substitution occurs. In general, this substitution is favorable when the incident atom is desired to dope

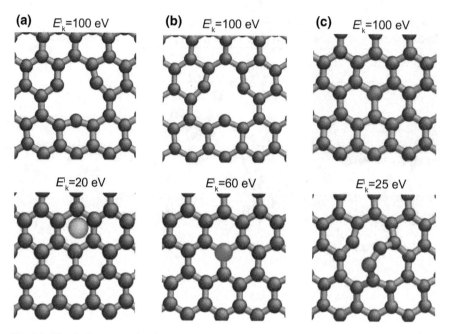

Fig. 5.4 The final patterns of defects under **a** Au, **b** Fe and **c** O bombardments at different E_k^i

into graphene [26]. Otherwise, it is considered as functionalized contamination [10]. In addition, this one-step substitution cannot occur for Au bombardments. Actually, after the head-to-head bombardment, the kinetic energy of incident atom E_k^i transfers to the kinetic of impacted carbon atoms E_k^C, bond energy E_b and its residual kinetic energy E_k^r. If $E_k^r < E_b$, the incident atom cannot escape from the graphene sheet and stay into the vacancy, and its residual kinetic energy eventually transfers to an increase of temperature. This results a one-step substitution. Since the Fe–C and O–C bonds are chemical bonds with high enough E_b, it is possible to satisfy the one-step substitution condition $E_k^r < E_b$. However, for Au bombardments, since the Au–C bond order is ∼0, it is impossible to satisfy the one-step substitution condition $E_k^r < 0$. Therefore, the one-step substitution is observed in Fe and O bombardments (Fig. 5.4b, c) but not in Au bombardments (Fig. 5.4a). In general, due to the physical interactions between Au and carbon atoms, the Au atom will either attach to the graphene sheet/bounce back at low E_k^i, or penetrate graphene sheet at high E_k^i, which is consistent with the experimental results [10].

The final pattern of defects after bombardments is determined by the kind of incident atom, impact site and the incident energy. Based on the effects of impact site and the physical properties of incident atoms, an energy spectrum of generating defects in graphene for O, Fe and Au bombardments at different E_k^i is summarized statistically (Fig. 5.5). As E_k^i increases, functionalized contamination, one-step substitution, pristine SV and random MV are generated sequentially. In addition, there exist a gap between one-step substitution and pristine SV. This is in good agreement

5.2 The Defect Controllability of Graphene Under Atomistic Bombardment

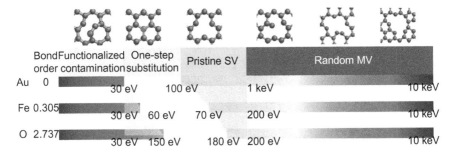

Fig. 5.5 Energy spectrum of defecting graphene for O, Fe and Au bombardments at different E_k^i

with experimental results [10] For the controllable defect technology, the one-step substitution and pristine SV are most concerned.

For the one-step substitution, the applicable ranges of E_k^i for Au, Fe and O are none, 30–60 eV, and 30–150 eV, respectively. As the bond order widens, the applicable ranges for the one-step substitution narrows, which is consistent with the residual kinetic energy analysis. For the pristine SV, the applicable ranges of E_k^i for Au, Fe and O are 0.1–1 keV, 70–200 eV, and 180–200 eV, respectively. As the bond order increases, the applicable ranges for the pristine SV narrow due to the atom-drag effect. According to this energy spectrum, choosing a proper incident atom with proper energy can highly improve the defect probability in graphene monolayer.

5.3 Graphene Nanoribbon with Grain Boundary Under Torsion Loading

Defects in graphene are equivalent to grain boundaries, which generally increase the discontinuous effects. According to the results in Chaps. 3 and 4, the discontinuous effects on the static and dynamic out-of-plane deformation of graphene are similar. Therefore, in this section, a representative case including both the static and dynamic out-of-plane deformation of graphene, the mechanical behavior of graphene nanoribbon (GNR) under torsion loading, is analyzed to reveal the effects of defects.

5.3.1 MD Simulation Details

The edge pattern of GNRs plays a dominant role in their electronic properties as well as mechanical behaviors [27]. Previous experimental and theoretical results show that the band gap of armchair GNRs (AGNRs) can be opened by strain, while zigzag GNRs (ZGNRs) remain metallic [28]. Therefore, considering the applicability in NEMS, the torsion behaviors of AGNRs and tilt-armchair-edge GNRs (tAGNRs)

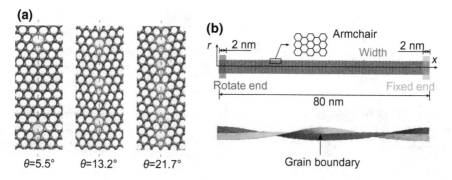

Fig. 5.6 (a) The grain boundaries with different mismatch angle θ; (b) The atomistic configurations of GNRs

are studied in this section. There are three types of tAGNRs with different mismatch angle $\theta = 5.5°$, $13.2°$ and $21.7°$ (Fig. 5.6a). The initial tAGNRs are modeled by C++ codes (Appendix B), and then relaxed through energy minimization simulation to reach a low energy state. A torsion loading is applied to the left end of a GNR with grain boundary along x-axis (the axle wire) (Fig. 5.6b). The length of GNR is 80 nm, and the width of ends is 2 nm.

MD simulations are performed by LAMMPS [24]. The interactions between carbon atoms are described by an adaptive intermolecular reactive empirical bond order (AIREBO) potential [29]. In addition, the first-principle calculation results of Jia et al. show that GNRs are stable with a minimum width of $\sim$1.0 nm [30]. Therefore, the minimum width of GNRs W_{min} is set to 1.0 nm. The time step is set to 1.0 fs for all MD simulations. Before torsion simulations, GNRs are relaxed at 300 K for 100 ps with NVT ensemble. Torsion simulations are performed with NVE ensemble, and the GNR is relaxed for 10 ps to reach equilibrium state every rotation of $1°$. Visualization is performed using VMD [31] and AtomEye [25]. As analyzed in Chap. 2, since the width of GNRs W is 1.0–10.0 nm, it is challenging to directly characterize the structural stress based on atomic stresses due to the fluctuations caused by fewer statistical samples. Therefore, the potential energy of GNRs is used to characterize the deformation and stress states.

In addition, the electron transfer properties of tAGNRs under torsion loading are briefly discussed. The current–voltage (I–V) characteristics are calculated using the open source code OpenMX [32]. For each bias voltage, the electronic structure of tAGNRs is self-consistently determined at an electronic temperature of 300 K by means of a non-equilibrium Green's function (NEGF) method [33] coupled with a local density approximation (LDA) [34]. Pseudoatomic orbital (PAO) basis functions of C-s^2p^2 and H-s^2 with a cutoff radius of 5.0 bohr are used [35], and the convergence criterion for total energy is set to 1.0×10^6 Hr. Realspace grid techniques are used with an energy cutoff of 150 Ry in numerical integration, and the Poisson equations are solved through a fast Fourier transformation.

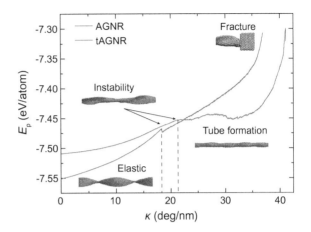

Fig. 5.7 The E_p–κ curves of common AGNR and tAGNR at $W = 2.5$ nm. The mismatch angle of tAGNR $\theta = 5.5°$

5.3.2 The Instability of GNRs Under Torsion Loading

The twist rate of GNRs is defined as:

$$\kappa = \frac{d\Phi}{dx}, \tag{5.1}$$

where Φ is the twist angle. If the deformation of GNRs is homogeneous, Eq. (5.1) can be expressed as:

$$\kappa = \frac{d\Phi}{dx} = \frac{\Phi}{L}, \tag{5.2}$$

where $L = 76$ nm is the mobile length of GNRs. In general, as κ increases, the average atomic potential energy E_p increases. According to the representative E_p–κ curves (Fig. 5.7), there are three stages:

- When κ is less than a critical value κ_c ($\kappa < \kappa_c$), the GNR is in the elastic state and twisted homogeneously.
- When $\kappa > \kappa_c$, instability of the GNR occurs, which leads to an inflection point on the E_p–κ curve. After the instability, the GNR immediately transforms into tubes, and E_p increases with κ.
- Edge fracture occurs due to a concentration of stress.

These stages are in good agreement with the results of Cranford and Kit et al. [36, 37]. Compared with the AGNRs, there exists a yield stage of E_p, which will be discussed in following subsections. In addition, the representative cases show that, the critical twist rate κ_c of tAGNR is higher than that of common AGNR at an identical width W.

The critical twist rate κ_c is important for the application of GNRs in NEMS [30], which can be considered as a function of W and θ: $\kappa_c = \kappa_c(\theta, W)$. κ_c decreases with the increase of W (Fig. 5.8a). To compare with the common AGNRs ($\theta = 0$), an

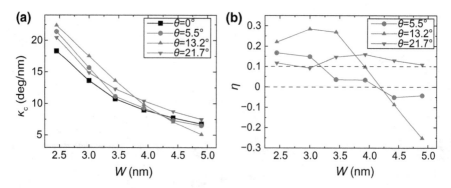

Fig. 5.8 The **a** $W-\kappa_c$ and **b** $W-\eta$ curves of tAGNRs at different mismatch angle θ. The common AGNRs are classified as $\theta = 0°$

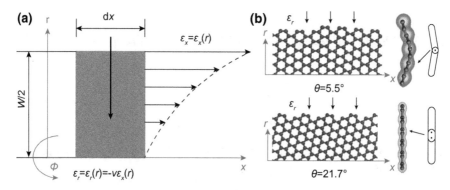

Fig. 5.9 **a** The stress status schematic of a differential dx in GNR; **b** The atomic electron density in flexural plane ($r-\Phi$ plane) at different mismatch angle θ

incremental factor η of κ_c is defined as:

$$\eta(\theta, W) = \frac{\kappa_c(\theta, W) - \kappa_c(0, W)}{\kappa_c(0, W)}, \tag{5.3}$$

where $\eta > 0$ indicates that the κ_c of tAGNR is higher than that of common AGNR. When $W < 4.0$ nm, κ_c of tAGNRs is generally 10% larger than that of common AGNRs except for the cases at $\theta = 5.5°$ (Fig. 5.8b); when $W > 4.0$ nm, κ_c of tAGNRs is generally less than that of common AGNRs except for the cases at $\theta = 21.7°$. Actually, the length of GNRs applied in NEMS is generally over 100 nm, while their width is less than 4.0 nm [38]. Therefore, the permissible torsion angle of tAGNRs is generally 100° higher than that of common AGNRs, which highly affects the electronic properties [30, 39].

In general, defects degrade the mechanical properties of materials. To clarify the anomalous torsion behavior of tAGNRs, the stress state of tAGNRs, which is x-axial symmetric, is analyzed through a differential dx with a half width $W/2$ in the

5.3 Graphene Nanoribbon with Grain Boundary Under Torsion Loading

x–r–Φ cylindrical coordinate system (Fig. 5.9a). The deformation ds along x-axis is a function of r:

$$\mathrm{d}s = \mathrm{d}s(r) = \sqrt{(\mathrm{d}x)^2 + [\mathrm{d}r\cos(\kappa x)]^2 + [\mathrm{d}r\sin(\kappa x)]^2} = \sqrt{1 + (\kappa r)^2}\,\mathrm{d}x. \quad (5.4)$$

Therefore, the tensile strain along x-axis is expressed as:

$$\varepsilon_x = \frac{\mathrm{d}s - \mathrm{d}x}{\mathrm{d}x} = \sqrt{1 + (\kappa r)^2} - 1. \quad (5.5)$$

Due to the Poisson's effect, the compressed strain along r-axis is given as:

$$\varepsilon_r = \varepsilon_r(r) = -\nu\varepsilon_x(r), \quad (5.6)$$

where ν is the Poisson's ratio of graphene. Since the ends of GNRs are fixed, this tensile-induced compression is similar to the buckling in an elastic sheet under tension loading [40]. As analyzed in Chap. 3, the critical strain for buckling in graphene monolayer is generally less than 0.02. Therefore, the strain–stress relation is considered linear [41]:

$$\sigma_r = -\nu E\varepsilon_x(r), \quad (5.7)$$

where E is the Young's modulus of graphene. Based on the angular momentum balance of the system, the maximum compression stress $\sigma_{r,\max}$ along r-axis is given as:

$$\sigma_{r,\max} = -\frac{2\nu E}{W}\int_0^{W/2} [\varepsilon_x(r) - \varepsilon_x(0)]\,\mathrm{d}r. \quad (5.8)$$

Assuming $F(r) = \int \varepsilon_x(r)\mathrm{d}r$, the Eq. (5.8) can be expressed as:

$$\sigma_{r,\max} = -\frac{2\nu E}{W}F(W/2) + \nu E\varepsilon_x(0), \quad (5.9)$$

which is a function of W: $\sigma_{r,\max} = \sigma_{r,\max}(W)$. The derivative of $\sigma_{r,\max}$ with respect to W is expressed as:

$$\begin{aligned}
\frac{\mathrm{d}\sigma_{r,\max}}{\mathrm{d}W} &= -2\nu E\frac{\mathrm{d}}{\mathrm{d}W}\left[\frac{F(W/2)}{W}\right] \\
&= -2\nu E\left[\frac{1}{2}F'(W/2)W - F(W/2)\right] \\
&= -2\nu E\left[\frac{1}{2}\varepsilon_x(W/2)W - F(W/2)\right] \\
&= -2\nu E\int_0^{W/2}[\varepsilon_x(W/2) - \varepsilon_x(r)]\,\mathrm{d}r < 0.
\end{aligned} \quad (5.10)$$

Therefore, the absolute value of $\sigma_{r,\max}$ increases with W. This explains that κ_c decreases with the increase of W (Fig. 5.8a).

In addition, as analyzed above, there exist two exceptions when $\theta = 5.5°$ and $21.7°$ (Fig. 5.8b). In general, the effect of defects is considered degrading the buckling strength of GNRs due to a high potential energy around the grain boundaries [37, 42]. However, as analyzed in Chaps. 3 and 4, the discontinuous effect plays a dominant role in the buckling of graphene when the compressed length is less than 2 nm, which is also chirality-dependent. When $W < 4$ nm, the compressed length of GNRs along r-aixs is less than 2 nm, which induces the discontinuous effect. For the common AGNRs, the compression direction is zigzag-along, which is easy to buckle. However, for the tAGNRs, the compression direction is rotated nearly from zigzag-along to armchair-along (Fig. 5.9b), which leads to an increase of the critical strain for buckling due to the discontinuous effect. For the tAGNRs with $\theta = 5.5°$, the κ_c at $W = 3.5$ nm and 4.0 nm are only ~3% higher than that of common AGNRs. Actually, when $\theta = 5.5°$, the compression direction in tAGNRs is close to that of common AGNRs, and the discontinuous effect is inconsiderable to increase κ_c. For the tAGNRs with $\theta = 21.7°$, the compression direction has been rotated to the direction between armchair- and zigzag-along, which induces a strong discontinuous effect as analyzed in Chaps. 3 and 4. Therefore, the κ_c of tAGNRs with $\theta = 21.7°$ is still about 10% higher than that of common AGNRs.

As mentioned above, compared with the common AGNRs, there is a yield stage for tAGNRs (Fig. 5.7). Actually, the buckling occurs in the middle of tAGNRs along x-axis, and the whole buckling process can be characterized by the deformation of the cross section (Fig. 5.10). There are two different points between AGNRs and tAGNRs. Firstly, there is a deflection along r-axis for tAGNRs but not for AGNRs, which results that $\varepsilon_x(0) = 0$ for AGNRs while $\varepsilon_x(0) > 0$ for tAGNRs. According to Eq. (5.9), at a same κ, the $\sigma_{r,\max}$ for AGNRs is lager than that of tAGNRs. However, when $W > 4$ nm, the discontinuous effect is negligible, while the κ_c of tAGNRs is still less than that of AGNRs. Therefore, compared with the discontinuous effects, the effect of deflection is negligible in the buckling. Secondly, a spontaneous creasing occurs at grain boundary in tAGNRs with a stable creased angle (Fig. 5.10b). During the buckling, this creased angle will be bent to a reflex angle, which indicates an incremental strain energy. Considering that the structural instability propagates from the middle of tAGNRs to the ends, all the creased angles along x-axis should be bent, which results in the yield stage.

5.3.3 The Electron Transport Properties of tAGNRs

In addition to the mechanical properties, the electronic properties of AGNRs can be changed by the defects [43]. The electron transport is one of the most important electronic properties of AGNRs [44]. Actually, the grain boundaries change the direction and local width of AGNRs, whose effects are similar to the junctions [45]. DFT simulations are performed to evaluate the electron transport properties of

5.3 Graphene Nanoribbon with Grain Boundary Under Torsion Loading

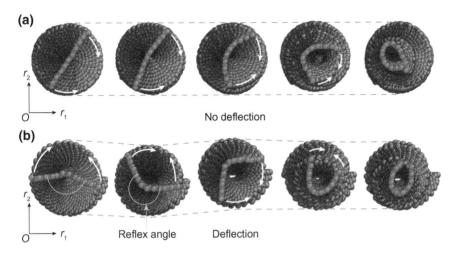

Fig. 5.10 The cross sections of **a** common AGNRs and **b** tAGNRs. The carbon atoms in the cross section are colored in red

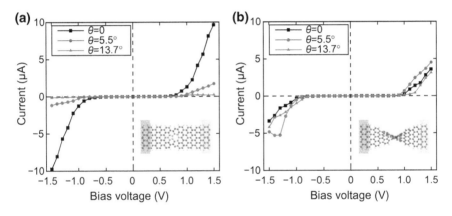

Fig. 5.11 The current–voltage curves of AGNRs and tAGNRs with twist angles of **a** 0° and **b** 180°

$3p$–AGNRs (here $p = 3$) [38] and tAGNRs, and hydrogen atoms are added on their edges to avoid dangling bonds [46, 47]. Two columns of carbon atoms (colored in shadows in Fig. 5.11) are set as electrodes. The electrical resistance of $3p$–AGNRs increases with the twist angle, and the band gap is not significantly affected, which is in good agreement with the results of Jia et al. [30]. However, for the tAGNRs, the band gap is ~0 under a twist angle of 0, while opened under a twist a twist angle of 180°. In addition, the current–voltage curves of tAGNRs are asymmetrical due to the asymmetrical defects along grain boundaries. Therefore, the defects in tAGNRs have a positive effect on the mechanical properties, as well as their electrical properties, which is important for the AGNR-based nanodevices in electromechanical systems.

5.4 Chapter Summary

In this chapter, the defect technology of graphene is reviewed and discussed, and the defect-induced discontinuous effects on the mechanical behaviors of graphene nanoribbon under torsion loading are studied. The basic conclusions are listed below:

- For the atom bombardments, as the electronegativity of incident atom increases, the applicable incident energy range for one-step substitution widens, while that for generating single-vacancy narrows.
- The grain boundary can increase the discontinuous effects in graphene, which leads to an anomalous twisting strength of tAGNRs when the width of GNR is less than 4.0 nm.
- When the width of GNR is over 4.0 nm, the discontinuous effects induced by grain boundary cannot lead to a strengthening effect, while deteriorate the twisting strength of tAGNRs due to the structural instability at the increased scale.

The out-of-plane mechanical behaviors of graphene are affected by the defect-induced discontinuous effects, which still follow the fundamental mechanisms raised in previous chapters.

References

1. Ferreira A, Xu X, Tan CL, Bae SK, Peres N, Hong BH, Özyilmaz B, Neto AC (2011) Europhys Lett 94(2):28003
2. Grantab R, Shenoy VB, Ruoff RS (2010) Science 330(6006):946
3. Carlsson JM, Scheffler M (2006) Phys Rev Lett 96(4):046806
4. Santos EJ, Sánchez-Portal D, Ayuela A (2010) Phys Rev B 81(12):125433
5. Khurana G, Kumar N, Kotnala R, Nautiyal T, Katiyar R (2013) Nanoscale 5(8):3346
6. Bell DC, Lemme MC, Stern LA, Williams JR, Marcus CM (2009) Nanotechnology 20(45):455301
7. Fischbein MD, Drndić M (2008) Appl Phys Lett 93(11):113107
8. Zhu W, Wang H, Yang W (2012) Nanoscale 4(15):4555
9. Lemme MC, Bell DC, Williams JR, Stern LA, Baugher BW, Jarillo-Herrero P, Marcus CM (2009) ACS Nano 3(9):2674
10. Wang H, Wang Q, Cheng Y, Li K, Yao Y, Zhang Q, Dong C, Wang P, Schwingenschlogl U, Yang W et al (2011) Nano Lett 12(1):141
11. Åhlgren E, Kotakoski J, Krasheninnikov A (2011) Phys Rev B 83(11):115424
12. Krasheninnikov A, Nordlund K (2010) J Appl Phys 107(7):3
13. Lehtinen O, Kotakoski J, Krasheninnikov A, Tolvanen A, Nordlund K, Keinonen J (2010) Phys Rev B 81(15):153401
14. Bubin S, Wang B, Pantelides S, Varga K (2012) Phys Rev B 85(23):235435
15. Gottstein G (2013) Physical foundations of materials science. Springer Science & Business Media
16. Wei Y, Wu J, Yin H, Shi X, Yang R, Dresselhaus M (2012) Nat Mater 11(9):759
17. Van Duin AC, Dasgupta S, Lorant F, Goddard WA (2001) J Phys Chem A 105(41):9396
18. Han SS, Kang JK, Lee HM, van Duin AC, Goddard WA III (2005) J Chem Phys 123(11):114703
19. Nielson KD, van Duin AC, Oxgaard J, Deng WQ, Goddard WA (2005) J Phys Chem A 109(3):493

References

20. Järrvi TT, van Duin AC, Nordlund K, Goddard WA III (2011) J Phys Chem A 115(37):10315
21. Keith JA, Fantauzzi D, Jacob T, van Duin AC (2010) Phys Rev B 81(23):235404
22. Aryanpour M, van Duin AC, Kubicki JD (2010) J Phys Chem A 114(21):6298
23. Chenoweth K, Van Duin AC, Goddard WA (2008) J Phys Chem A 112(5):1040
24. Plimpton S (1995) J Comput Phys 117(1):1
25. Li J (2003) Modell Simul Mater Sci Eng 11(2):173
26. Krasheninnikov A, Lehtinen P, Foster AS, Pyykkö P, Nieminen RM (2009) Phys Rev Lett 102(12):126807
27. Gunlycke D, Li J, Mintmire JW, White CT (2010) Nano Lett 10(9):3638
28. Li Y, Jiang X, Liu Z, Liu Z (2010) Nano Res 3(8):545
29. Stuart SJ, Tutein AB, Harrison JA (2000) J Chem Phys 112(14):6472
30. Jia J, Shi D, Feng X, Chen G (2014) Carbon 76:54
31. Humphrey W, Dalke A, Schulten K (1996) J Mol Gr 14(1):33
32. Boker S, Neale M, Maes H, Wilde M, Spiegel M, Brick T, Spies J, Estabrook R, Kenny S, Bates T et al (2011) Psychometrika 76(2):306
33. Caroli C, Combescot R, Nozieres P, Saint-James D (1971) J Phys C: Solid State Phys 4(8):916
34. Ceperley DM, Alder B (1980) Phys Rev Lett 45(7):566
35. Ozaki T (2003) Phys Rev B 67(15):155108
36. Cranford S, Buehler MJ (2011) Modell Simul Mater Sci Eng 19(5):054003
37. Kit O, Tallinen T, Mahadevan L, Timonen J, Koskinen P (2012) Phys Rev B 85(8):085428
38. Son YW, Cohen ML, Louie SG (2006) Phys Rev Lett 97(21):216803
39. Sadrzadeh A, Hua M, Yakobson BI (2011) Appl Phys Lett 99(1):013102
40. Cerda E, Ravi-Chandar K, Mahadevan L (2002) Nature 419(6907):579
41. Zhao H, Min K, Aluru N (2009) Nano Lett 9(8):3012
42. Yi L, Yin Z, Zhang Y, Chang T (2013) Carbon 51:373
43. Mucciolo ER, Neto AC, Lewenkopf CH (2009) Phys Rev B 79(7):075407
44. Zou D, Cui B, Kong X, Zhao W, Zhao J, Liu D (2015) Phys Chem Chem Phys 17(17):11292
45. Motta C, Sánchez-Portal D, Trioni M (2012) Phys Chem Chem Phys 14(30):10683
46. Moraes Diniz E (2014) Appl Phys Lett 104(8):083119
47. Lin YT, Chung HC, Yang PH, Lin SY, Lin MF (2015) Phys Chem Chem Phys 17(25):16545

Chapter 6
Mechanical Behaviors of Graphene Nanolayered Composites

Graphene/matrix interface plays a dominant role in the mechanical properties of graphene composites. Based on the understanding of out-of-plane mechanical behaviors raised in previous chapters, the mechanical properties of graphene nanolayered composites are studied under two representative loads: out-of-plane shock and in-plane shear. The effects of graphene interfaces on the mechanical properties are studied, which is important for the design of graphene composites.

6.1 Introduction

As mentioned in Chap. 1, graphene is a 2D material with a strength of 130 GPa and a Young's modulus of 1 TPa [1, 2], which is considered as an ideal enhancer for composites [3]. The strength and toughness of graphene composites can be improved by specified microstructures [4]. Graphene forms a variety of interfaces, and these interfaces can block dislocation, which plays a dominant role in the strength of composites [5]. The results of Walker et al. show that the strong sp^2-bonded structure of graphene can heal the polymer [6], and the strengthening effects of graphene interfaces under in-plane loading are in good agreement with the classical theory of slip resistance [7]. Random doping of graphene into a matrix is a common and simple approach to fabricate graphene composites with randomly distributed graphene interfaces [8]. However, for these composites, even if the proportion of doped graphene is identical, different microstructures result in different mechanical properties [9]. In 2013, Kim et al. realized the synthesis of graphene–metal (copper and nickel) nanolayered (GMNL) composites [10], and these composites with ordered graphene/metal interfaces exhibit strengthening, toughening, anti-radiation and self-healing effects under different loading conditions [11]. In addition, compared with the randomly doped graphene composites, the study of the nanolayered composites can gain a deeper and clearer understanding of interfacial effects, which is important for further design of graphene composites and other nanolayered materials [12]. Actually, the mechanical behaviors of graphene nanolayered composites can also be classified as in-plane and out-of-plane due to the existence of graphene interfaces. To reveal the strengthening effects of graphene interfaces in composites, two representative loads, out-of-plane shock and in-plan shear, are studied in this chapter.

The out-of-plane loading is a simple and common method to result in out-of-plane deformation of graphene. The bending rigidity of graphene monolayer is only about 1.44 eV [13], and in theory the graphene interfaces are quite weak under out-of-plane loading. However, according to the research of weakening shock waves by weak interfaces, there may also exist strengthening effects in graphene-metal nanolayered composites [14]. A paradox to strengthen the composites under shock loading is that the interfaces should be weak and strong simultaneously [15]. On one hand, weak interfaces can weaken shock waves due to their high resistance, while they are generally destroyed after shock (e.g., Kurdjumov–Sachs interface) [16]. On the other hand, strong interfaces can block the dislocations and heal the materials, while their low resistance results in a coherent propagation of shock waves [17, 18]. For GMNL composites, graphene is difficult to be destroyed, and the interfacial interaction between graphene and metal is relatively weak, which may solve the paradox of interface under shock loading.

As analyzed in Chap. 3, buckling can occur at a compressive strain of $\sim$0.01. Under the complex stress states with an in-plane compressive component, the mechanical behaviors of graphene interfaces are important for the structural responses of composites, which has rarely been reported [19]. The mechanical behaviors of the GMNL composites under in-plane shear loading is representative to reveal the strengthening effects, which is also one of the complex stress state mentioned above [20, 21]. Although the nominal in-plane shear strength of graphene (atoms are constrained along the out-of-plane direction) is over 50 GPa, the actual shear strength is tremendously deteriorated due to the structural buckling [22], and the high nonlinearity induced by buckling limits its applications as structural materials [23]. The chirality-dependent mechanical behaviors of graphene discussed in previous chapters are also crucial in the design of anisotropic nanomaterials [24]. In addition, in the graphene-metal nanolayered composites, the thickness of metal is at nanoscale, whose shear strength is close to its theoretical upper limitation [25–27]. Moreover, the results of Mara et al. show that that interfaces in nanolayered composites can trap dislocations [28]. Therefore, it is expected that graphene interfaces may lead to strengthening and self-healing effects for GMNL composites under shear loading.

6.2 The Strong/Weak Duality of Graphene Interfaces Under Out-of-Plane Shock Loading

In this section, the mechanical behaviors of graphene-metal nanolayered (GMNL) composites under out-of-plane shock loading are studied to reveal the strengthening effects of graphene interfaces.

6.2.1 Models and Methods

A bullet-target model (Fig. 6.1a) and 1D shock model (Fig. 6.1b) are simulated to study the propagation of shock waves in GMNL composites. For the bullet-target model, the bullet is a metallic cylinder with a diameter of 2 nm, and the target is a plate of GMNL composites with two metallic layers and a graphene interface. The diameter and thickness of the target is 20 nm and 6 nm, respectively. The bullet is aiming at the center of the target with an incident velocity. For the 1D shock model, periodic boundary conditions are applied along y- and z-axes, and the y and z dimensions of simulation box are compatible with both the metal and graphene lattice simultaneously. The shock loading is applied along x-axis by assigning an initial particle velocity u_p of the whole configurations. The repeat layer spacing λ of GMNL composites is ranging from 1.4 to 8 nm, and the u_p is ranging from 0.1 to 1.0 km/s. Both copper- and nickel-matrix graphene nanolayered (GCuNL and GNiNL) composites are simulated according to previous experiments [10].

The molecular dynamics (MD) simulations are performed by open source codes LAMMPS [29]. The adaptive intermolecular reactive empirical bond order potential [30], and the embedded atom model (EAM) potentials [31], the interaction between metal atoms. The carbon–copper and carbon–nickel interatomic interactions are described by Lennard–Jones potentials [32, 33]. All initial configurations

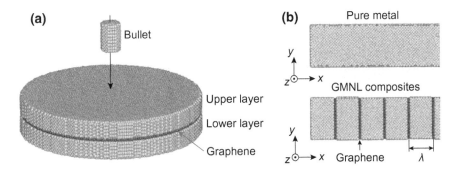

Fig. 6.1 a The schematic of bullet-target model; b Partial configurations of pure metal and GCuNL composites for 1D shock simulations. Shock loading is applied along x-axis

are relaxed with NPT ensemble at 300 K for 100 ps, and the shock simulations are performed with NVE ensemble. Visualization is performed using AtomEye [34].

6.2.2 The Strengthening Effects of Graphene Under Bullet Impact

Actually, the bullet-target model involves more than the propagation of shock waves in GMNL composites (e.g., the energy delocalization [35]), while is the most intuitive performance of strengthening effects. To simplify the discussion, only the cases at the incident velocity of 6.0 km/s are simulated for the comparison between pure metal and GMNL composites.

MD results show that, the pure copper target is penetrated by the bullet, while the GCuNL composites target is not penetrated. After the collisions, the atomistic configurations are relaxed with NVE ensemble for 100 ps. For the pure copper target, the configuration is destroyed permanently with a hole (Fig. 6.2a). For the GCuNL composites target, although the upper layer is penetrated, the graphene and the lower layer is not penetrated (Fig. 6.2b). Melting occurs in the pure copper target and the upper layer of GCuNL composites target, but not in the lower layer. During the initial relaxation (time < 60 ps), the average temperature of the upper layer is remarkably higher than that of the lower layer (Fig. 6.2c), and the temperature difference decreases due to the thermal conduction, which cools the upper layer rapidly. The non-melting effect is also one of the important factors of the protective ability for materials [36, 37].

Similar phenomenon is observed for the GNiNL composites target, which may suggest a universal strengthening effect of metal composites with graphene interface.

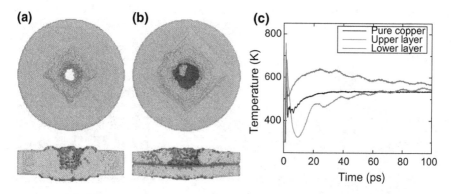

Fig. 6.2 **a** The pure copper target is penetrated. **b** The GCuNL composites target is not penetrated. **c** The temperature–time curves of copper for different targets. Different coloration of atoms represents different coordination numbers, which is calculated by AtomEye [34]

6.2.3 The Effects of Graphene Interface on Shock Waves

The 1D shock loading simulations of pure copper and GCuNL composites are performed to clarify the strengthening effects of graphene interfaces on the shock waves [38]. The crystal orientation of copper layers in GCuNL composites is identical with that of pure metal for comparison. The GCuNL composites are relaxed before shock simulations to obtain a proper stress state for graphene–copper interfaces. During the propagation of shock waves, the slip bands are stable and continuous, which is generated in the (111) plane (Fig. 6.3a). However, in the GCuNL composites, the slip bands are blocked by the graphene interfaces, and they are unstable with interlayer reflections (Fig. 6.3b). Since the damage caused by shock waves is mainly shear failure, graphene interfaces are difficult to be destroyed due to the sp^2-bonded structures. In addition, the interactions between graphene and copper are mainly the Van der Waals forces, which are much weaker than the stress along x-axis. Therefore, the graphene interfaces can be approximately considered as a free boundary along x-axis, and this boundary can impede the interfacial instabilities. Although the mechanical properties of graphene are strongly nonlinear under large deformation [2, 39], the real strain under the 1D shock loading is quite small due to the effect of free boundary. In this situation, the nonlinear behavior of graphene interface does not affect the propagation of shock waves, which is also consistent with the weak performance of Kurdjumov–Sachs interface [15, 16]. The nucleation of dislocation is impeded by the real-time self-healing during the propagation of shock waves, which is also consistent with the results of bullet-target simulations.

The stress states under shock loading can be analyzed by the Binning-analysis methods as introduced in Chap. 2. The crystal orientation of metal cannot weaken the shock waves [40], which is also consistent with the results of MD simulations in this section. Therefore, the following discussion is mainly focus on the mechanical response of pure copper and GMNL composites under shock loading along [100] orientation. Compared with the pure copper, there exist two additional regimes in the x–t diagram of stress σ_{xx} (decompressed and healing regimes marked in Fig. 6.4b): in the decompressed regime, the amplitude of σ_{xx} decreases; in the healing regime, the slip bands in copper layers disappear with the increase of temperature, which can be considered as a self-healing process. In addition, the melting, which is a damage under high-pressure shock loading [36], is impeded by the graphene interfaces. The results of GNiNL composites are similar to that of GCuNL composites.

The stress σ_{xx} in decompressed regime can be expressed as:

$$\sigma_{xx}^d = (1 - \eta)\sigma_{xx,0}, \qquad (6.1)$$

where η is defined as a damping factor of stress, $\sigma_{xx,0}$ is the stress in the shocked regime. The jump stress $[\sigma_{xx}]$ between the shocked and decompressed regimes is given as:

$$[\sigma_{xx}] = \sigma_{xx,0} - \sigma_{xx}^d = \eta\sigma_{xx,0}. \qquad (6.2)$$

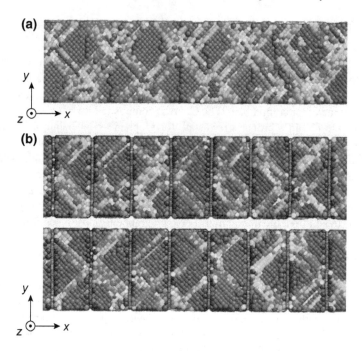

Fig. 6.3 a Slip bands are stable in pure copper; **b** Slip bands are unstable and can be healed in GCuNL composites. The shock loading is applied along [100] orientation, and the coloration of atoms represents different coordination number to show the slip bands

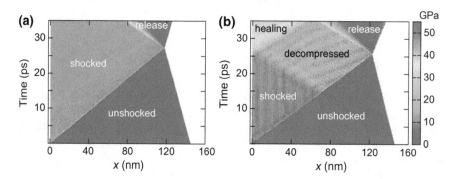

Fig. 6.4 The x–t diagrams of stress σ_{xx} for **a** pure copper and **b** GCuNL composites at $u_p = 0.5$ km/s, $\lambda = 3.0$ nm

The Rankine–Hugoniot relationship for the 1D shock includes:

$$\begin{cases} \rho_0 u_s = \rho(u_s - u_p) \\ \sigma_{xx} = \rho u_s u_p \end{cases}, \qquad (6.3)$$

6.2 The Strong/Weak Duality of Graphene Interfaces ...

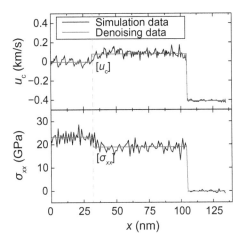

Fig. 6.5 Profiles of u_c and σ_{xx} along x-axis at $t = 23$ ps, $u_p = 0.5$ km/s, $\lambda = 3.0$ nm

where ρ and u_s are the density of the material and the shock wave speed, respectively. Substitution of Eq. (6.3) into Eq. (6.2) leads to the expression of $[\sigma_{xx}]$ as

$$[\sigma_{xx}] = \frac{\rho_0 u_s^2}{u_s - u_p}[u_p] = \eta \rho_0 u_s u_{p,0}. \tag{6.4}$$

Actually, the term of the center-of-mass velocity u_c should be removed during calculating the particle velocity u_p. Therefore, the jump of the center-of-mass velocity $[u_c]$ can also lead to a jump of u_p. MD results show that, the $[u_c]$ and $[\sigma_{xx}]$ occur at an identical x (Fig. 6.5). According to Eq. (6.4), the damping factor η is given as:

$$\eta = \frac{u_s}{u_s - u_p} \frac{[u_c]}{u_{p,0}}. \tag{6.5}$$

For the GCuNL composites, the effect of free boundary will result a reflection of velocity in the copper layers, which generates a center-of-mass velocity. Comparing with the value of $[\sigma_{xx}]/\sigma_{xx}$ and the result from Eq. (6.5), the relative error is less than 5%. Therefore, the decompressed regime is induced by the jump of u_c. In addition, the real-time self-healing can also partly recovery the elasticity of copper layers with a release of elastic energy, which also leads to a decrease of σ_{xx} [41].

When $\lambda > 2.0$ nm, The damping factor of stress η can be approximately fitted by a power function (Fig. 6.6):

$$\eta = \alpha u_p^\beta, \tag{6.6}$$

where α and β are two parameters. The parameter α is inversely proportional to λ, and β is determined by the $[u_c]$–u_p relation (Eq. (6.5)). However, when $\lambda < 2.0$ nm, the interaction between the adjacent graphene interfaces cannot be neglected [42],

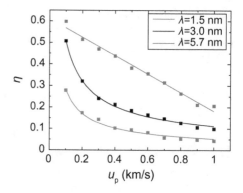

Fig. 6.6 u_p–η curves at different λ

which leads to a linear relation of u_p–η (Fig. 6.6). In general, the damping factor of stress η increases with the decrease of λ, which indicates that the resistance of shock waves increases.

6.3 The Mechanical Properties of GCuNL Composites Under In-Plane Shear Loading

In this section, the mechanical properties of GCuNL composites under in-plane shear loading are studied to reveal the effects of out-of-plane deformation of graphene.

6.3.1 Models and Methods

The ideal shear strength of pure copper is the minimum shear stress to cause a non-reversible deformation, which is generally the shear strength along $[11\bar{2}]$ direction in (111) plane [26]. At nanoscale, the strength of materials can be close to their theoretical upper limitation, which is highly influenced by the crystal orientation [43]. Since the repeat layer spacing of GCuNL composites is at nanoscale [10], the shear behaviors in the strongest (100) and the weakest (111) planes are both considered in the following discussion (Fig. 6.7a). For the GCuNL composites, the repeat layer spacing λ should be compatible with the copper lattice (Fig. 6.7b), and the dimensions of simulation box in xy plane should be compatible with the copper and graphene lattice simultaneously (Fig. 6.7c). In addition, the chirality of graphene is also important for the mechanical properties of GCuNL composites [22]. Therefore, four types of GCuNL composites with different interface configurations are investigated, which are simply referred to as (100)-armchair, (100)-zigzag, (111)-armchair and (111)-zigzag in the following discussion.

6.3 The Mechanical Properties of GCuNL Composites ...

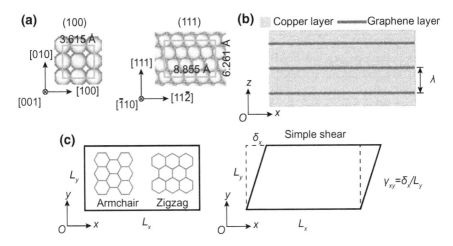

Fig. 6.7 **a** The copper lattice parameters in (100) and (111) planes; **b** The out-of-plane direction is along z-axis, and the repeat layer spacing λ should be compatible with the copper lattice; **c** The schematic of shear model for graphene, copper and GCuNL composites

MD simulations are performed using open source codes LAMMPS [29]. The interactions between carbon atoms are described by AIREBO potential [30] with a cutoff of 2 Å for the reactive empirical bond order (REBO) part [44], and that between copper atoms are described by the EAM potential [31]. A Lennard–Jones potential with parameters of 0.019996 eV and 3.225 Å is used to describe the van der Waals forces between carbon and copper atoms, which have been verified in previous research [11]. For all configurations, periodic boundary conditions are applied along x-, y- and z-axes. Before shear simulations, the atomistic configurations are relaxed with NPT ensemble for 100 ps at 300 K. The time step for all MD simulations is set to 1.0 fs. The mechanical properties are calculated by implementing the deformation-control method (Fig. 6.7c) [27]. The certain mechanical properties of graphene are not sensitive to the strain rate [45]. However, the deformation of copper depends on the strain rate owing to dislocation initiation [46]. At nanoscale, the process of dislocation propagation is independent of the shear strain rate at time scales sufficiently short to neglect creep and yet sufficiently long with respect to the sound speed of sound for strain rates below 0.0015 ps^{-1} [47, 48]. Therefore, a strain increment with strain rate of 0.001 ps^{-1} after every 1000 time steps with NVT ensemble at 300 K are applied in shear simulations [22, 27]. The shear strain γ is calculated as $\gamma = \gamma_{xy} = \delta_x/L_y$, where δ_x and L_y are the displacement along x-axis and the dimension of simulation box along y-axis, respectively. Visualization and dislocation analysis are performed using OVITO [49].

6.3.2 Shear Responses of GCuNL Composites

For the pure copper, the slip bands are generated in (111) planes under shear loading (Fig. 6.8a), and the shear yield stresses are 10.0 and 2.6 GPa for the (100)- and (111)-stacking copper, respectively. This is consistent with the results of Iskandarov et al. [27]. When the shear strain γ_{xy} is over the shear yield strain, plastic flow occurs in the (100)-stacking copper with a reduced shear stress of ~1.0 GPa, while τ of (111)-stacking copper periodically increases and decreases with generating more slip bands. For pure graphene, the theoretical and real shear strength is obtained by applying a periodic and a shrink-wrapped boundary condition along the out-of-plane direction, respectively [22]. The thickness of graphene monolayer h^G is assumed as 0.335 nm [50]. Since the graphene monolayer is fixed along the out-of-plane direction, its theoretical shear strength is over 50 GPa under both armchair- and zigzag-along shear loading (Fig. 6.8b). However, as discussed in Chaps. 3 and 4, structural wrinkles are generated due to the in-plane buckling, which deteriorates the shear strength. Actually, the buckled graphene monolayer cannot be used as structural materials due to the strong nonlinear mechanical behaviors [23]. In general, the out-of-plane deformation of graphene degrades its in-plane shear strength. Therefore, for GCuNL composites, if the out-of-plane deformation of graphene is impeded, the shear strength can be improved.

For the GCuNL composites, there exist three stages (elastic, yield and failure) under different shear strain, which are divided by the shear yield strain γ_Y and shear failure strain γ_F (Fig. 6.9). At the elastic stage ($\gamma < \gamma_Y$), there are not any dislocations in copper and graphene layers, and τ is approximately proportional to γ: $\tau = G\gamma$, where G is the shear modulus of GCuNL composites. At the yield stage ($\gamma_Y < \gamma < \gamma_F$), τ is approximately a constant due to the plastic flow of copper layers, and there are still not any fractures in graphene layers. At the failure stage ($\gamma > \gamma_F$), τ decreases rapidly at γ_F with the structural fractures in graphene layers.

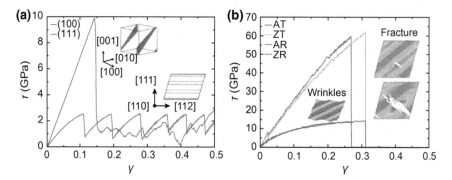

Fig. 6.8 **a** The γ–τ curves of (100)- and (111)-stacking copper; **b** The γ–τ curves of graphene monolayer. AT, ZT, AR and ZR represent the armchair-along-theoretical, zigzag-along-theoretical, armchair-along-real, and zigzag-along-real shear stress, respectively

6.3 The Mechanical Properties of GCuNL Composites ...

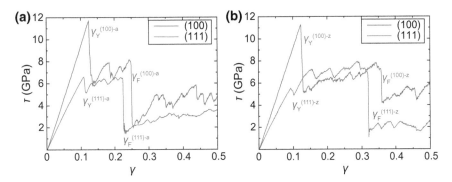

Fig. 6.9 Representative $\gamma-\tau$ curves of **a** armchair- and **b** zigzag-stacking GCuNL composites at $\lambda = 4.0$ nm

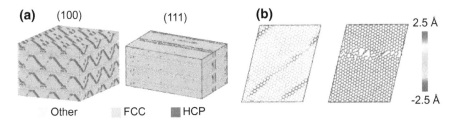

Fig. 6.10 a Slip bands generated in copper; **b** The fracture of graphene in GCuNL composites

Compared with pure graphene [22] and copper [27], the mechanical properties of GCuNL composites under shear loading are remarkably influenced by the chirality of graphene and the crystal orientation of copper.

MD results show that, the initial plasticity occurs with the slip bands generated in (111) planes for both (100)- and (111)-stacking GCuNL composites (Fig. 6.10a). Since the out-of-plane deformation of graphene is limited by the copper layers (the maximum out-of-plane displacement is generally less than 3.0 Å), structural buckling does not occur in GCuNL composites (Fig. 6.10b). Therefore, the shear strength of graphene in GCuNL composites is close to that under theoretical in-plane shear loading [22], and its fracture occurs without wrinkles (Fig. 6.10b).

6.3.3 The Yield and Failure of GCuNL Composites

According to the results of Sect. 6.2, the interfacial interaction between graphene and copper is relatively weak compared with the in-plane stresses. Therefore, the shear stress τ at elastic stage can be expressed as the weighted average of stresses in the graphene (τ^G) and copper (τ^{Cu}) layers:

$$\tau = \tau^G \eta^G + \tau^{Cu} \eta^{Cu} = G^G \gamma \eta^G + G^{Cu} \gamma \eta^{Cu}, \tag{6.7}$$

where G^G and G^{Cu} the shear modulus of graphene and copper, respectively. η^G and η^{Cu} are the volume fractions of graphene and copper in GCuNL composites, which can be expressed as:

$$\begin{cases} \eta^G = h^G/\lambda \\ \eta^{Cu} = 1 - \eta^G = 1 - h^G/\lambda \end{cases}. \tag{6.8}$$

According the analysis above, the shear yield strain of GCuNL composites γ_Y is theoretically a constant equal to that of pure copper, which is influenced by the crystal orientation of copper. The shear yield stress of GCuNL composites τ_Y can be expressed as:

$$\tau_Y = G^{Cu} \gamma_Y + (G^G - G^{Cu}) \gamma_Y \frac{h^G}{\lambda}. \tag{6.9}$$

MD results show that, the λ–γ_Y (Fig. 6.11a) and λ–τ_Y (Fig. 6.11b) relations are in good agreement with the theoretical predictions based on Eq. (6.9) except for (100)-stacking GCuNL composites at $\lambda < 3.0$ nm. Due to the out-of-plane attractive forces between graphene and copper layers [51, 52], the small wrinkles of graphene (Fig. 6.10b) lead to additional in-plane stresses in copper layers. Therefore, the γ_Y is generally 5% less than that of pure copper at an identical crystal orientation (Fig. 6.11a). Under shear deformation, there exists an interfacial shear stress τ_i due to the friction [53, 54]:

$$\tau_i = \frac{f_i S}{\lambda S} = \frac{f_i}{\lambda}, \tag{6.10}$$

where S is the area of graphene interface, f_i is the line density of interfacial force along out-of-plane direction. Since the graphene only interacts with the copper atoms within a certain thickness near the interface, f_i can be considered as a constant [55]. Therefore, τ_i increases with the decrease of λ. For the (100)-stacking GCuNL composites, the resultant stress of interfacial shear and tension stress can induce oblique slip bands in (100)-stacking GCuNL composites at $\lambda < 3.0$ nm (Fig. 6.11c). For (111)-stacking GCuNL composites, this effect is negligible due to the slip bands perpendicular to the interface.

At the yield stage, the shear stress of GCuNL composites can be expressed as:

$$\tau = \tau^G \eta^G + \overline{\tau}^{Cu} \eta^{Cu}, \tag{6.11}$$

where $\overline{\tau}^{Cu}$ is the plastic flow stress of copper, which is generally a constant. According to Eq. (6.7), the γ_F is determined by the fracture of graphene in GCuNL composites, which is chirality-dependent. Compared with the theoretical shear failure strain along armchair-($\gamma_F^{armchair}$) and zigzag-(γ_F^{zigzag}) directions (Fig. 6.12a), MD results are also constants but generally 10% less than them. Although the structural buckling of graphene is impeded in GCuNL composites, theres exit some small wrinkles (the insets in Fig. 6.12a), which decreases the shear strength of graphene. Substitution of

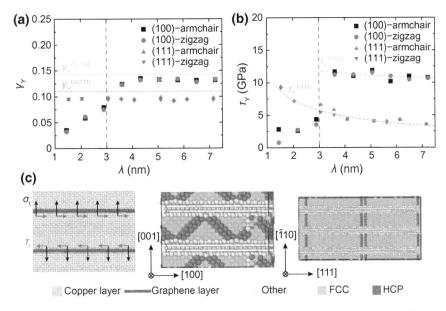

Fig. 6.11 The **a** λ–γ_Y and **b** λ–τ_Y relations are in good agreement with the theoretical predictions except for (100)-stacking GCuNL composites at $\lambda < 3.0$ nm. **c** The schematic of interfacial stresses in GCuNL composites, and the dislocations in (100)- and (111)-stacking GCuNL composites

Eq. (6.8) into Eq. (6.11) leads to the expression of τ_F as:

$$\tau_F = \tau^G \eta^G + \overline{\tau}^{Cu} \eta^{Cu} = \frac{\tau^G h^G}{\lambda} + \frac{\overline{\tau}^{Cu}(\lambda - h^G)}{\lambda}. \qquad (6.12)$$

MD results are in good agreement with the theoretical predictions (Fig. 6.12b). In addition, τ_F of (100)-stacking GCuNL composites is remarkably higher than that of (111)-stacking. The slip bands in (100)-stacking GCuNL composites are rotated during the plastic flow (Fig. 6.12c), while this rotation does not occur in (111)-stacking GCuNL composites (Fig. 6.12d). This rotation results an increase of plastic strain energy, which eventually results the increase of τ_F.

6.3.4 Self-Healing Effects in GCuNL Composites

The in-plane sp^2 structures of graphene can block the dislocations near the interface, which eventually results a self-healing effect in composites [6]. A series of load–release simulations are performed to reveal the self-healing effects in GCuNL composites under shear loading, and the shear loading is released at $\gamma = 0.2$. MD results show that the self-healing processes in (100)- and (111)-stacking GCuNL composites are different: in the (100)-stacking composites (Fig. 6.13a), the adjacent

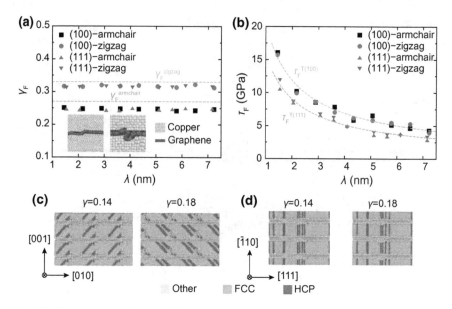

Fig. 6.12 The **a** λ–γ_F and **b** λ–τ_F relations are in good agreement with the theoretical predictions. The dislocations in **a** (100)- and **b** (111)-stacking GCuNL composites

slip bands are merged, then dragged to the interface, and eventually healed from the inner of copper layers to the interface; in the (111)-stacking GCuNL composites (Fig. 6.13b), although the adjacent slip bands are also merged, the dislocations are healed from the interface to the inner of copper layers. Actually, the healing can be considered as a reversed loading process, which is resulted by the interfacial stresses (Fig. 6.11c). Due to the tension asymmetry of copper [56, 57], the tensile stresses required for homogeneous dislocation nucleation in [100] and [110] orientations are about 9 GPa and 4 GPa, respectively. MD simulations show that the interfacial attractive stress is 2.5–5.0 GPa, which is consistent with the results of Muller et al. [11]. For (100)-stacking GCuNL composites, the dislocations in copper layers can only be healed by the resultant force provided by the neighbor graphene interfaces, which leads to the initial healing in the inner of copper layers. However, for (111)-stacking GCuNL composites, the interfacial stress is large than the tensile stress of dislocation nucleation, which can heal the dislocations near the interface directly.

To analyze the self-healing effect quantitatively, a self-healing ratio Π of GCuNL composites is defined as:

$$\Pi = \frac{\pi^{Cu} - \pi^{Comp}}{\pi^{Cu}}, \tag{6.13}$$

where π^{Cu} and π^{Comp} are the atomistic percentages of residual dislocations in pure copper and GCuNL composites after release, respectively. Π represents the self-healing ability of GCuNL composites relative to copper: $\Pi = 1.0$ indicates that all dislocations are healed, while a negative Π indicates an increase of dislocations.

6.3 The Mechanical Properties of GCuNL Composites ...

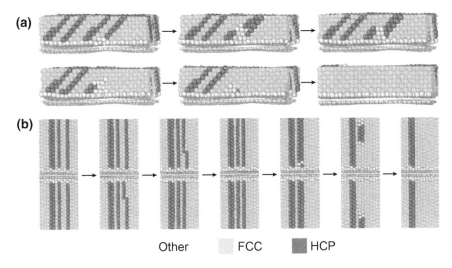

Fig. 6.13 The self-healing in **a** (100)- and **b** (111)-stacking GCuNL composites

According to MD results, π^{Cu} depends on the crystal orientation of copper, which is 5.6 at.% and 7.4 at.% for (100)- and (111)-stacking copper, respectively. In addition, the effects of chirality of graphene on the self-healing process are negligible. At $\lambda < 5.0$ nm, the Π of (100)-stacking GCuNL composites is generally less than 0 (Fig. 6.14a), while the Π of (111)-stacking fluctuates drastically (Fig. 6.14b). Actually, according to the interfacial stress analysis (Fig. 6.11c), the interfacial stress increases with the decrease of λ, which induces oblique dislocations in (111) planes. In addition, the interlayer tensile stress increases, which leads to an increase of the potential energy of copper layers and thermodynamical fluctuations during the self-healing process. On the other hand, when λ is over a critical value, Π is less than 0.05, which indicates approximate an identical self-healing ability between GCuNL

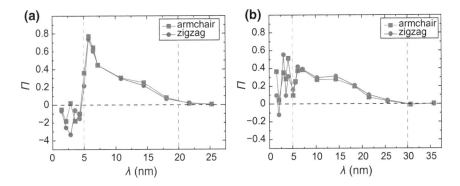

Fig. 6.14 The Π–λ curves of **a** (100)- and **b** (111)-stacking GCuNL composites

composites and pure copper. The critical λ for (100)- and (111)-stacking GCuNL composites are 20 nm and 25 nm, respectively. Considering the copper layers are generally polycrystalline [10], the upper limitation of λ for GCuNL composites with positive self-healing effect is about 20 nm. Therefore, considering both the strengthening and self-healing effect, the optimum value of λ for GCuNL composites is 5–20 nm.

6.4 Chapter Summary

In this chapter, the out-of-plane mechanical behaviors of graphene in nanolayered composites are studied. The basic conclusions are listed below:

- The out-of-plane deformation of graphene interfaces in nanolayered composites will deteriorate the in-plane strength of composites, while benefit the out-of-plane strength.
- During the propagation of shock waves, the graphene interface can be considered as a "free boundary", while this "free boundary" can remain intact and will not be destroyed.
- In general, the strength of GMNL composites increases with the decrease of repeat layer spacing. However, when the repeat layer spacing is less than a critical value, structural wrinkles will deteriorate the mechanical performance of composites. Therefore, there exists an optimum range of repeat layer spacing for GMNL composites.
- Graphene interface can impede the propagation of dislocations. In addition, graphene interface can induce a self-healing effect along the out-of-plane direction for GMNL composites, which is realized by the in-plane lattice constraint.

The interfacial interaction plays a dominant role in the mechanical properties of graphene–metal nanolayered composites, which can also provide some valuable guides for the design of other materials with similar structures.

References

1. Geim AK (2009) Science 324(5934):1530
2. Lee C, Wei X, Kysar JW, Hone J (2008) Science 321(5887):385
3. Huang X, Qi X, Boey F, Zhang H (2012) Chem Soc Rev 41(2):666
4. Rafiee MA, Rafiee J, Wang Z, Song H, Yu ZZ, Koratkar N (2009) ACS Nano 3(12):3884
5. Wang J, Hoagland R, Hirth J, Misra A (2008) Acta Materialia 56(19):5685
6. Walker LS, Marotto VR, Rafiee MA, Koratkar N, Corral EL (2011) Acs Nano 5(4):3182
7. Hoagland RG, Kurtz RJ, Henager C Jr (2004) Scripta Materialia 50(6):775
8. Hou G, Tao W, Liu J, Gao Y, Zhang L, Li Y (2017) Phys Chem Chem Phys 19(47):32024
9. Kuilla T, Bhadra S, Yao D, Kim NH, Bose S, Lee JH (2010) Progr Polymer Sci 35(11):1350
10. Kim Y, Lee J, Yeom MS, Shin JW, Kim H, Cui Y, Kysar JW, Hone J, Jung Y, Jeon S et al (2013) Nat Commun 4:3114

References

11. Muller SE, Nair AK (2016) JOM 68(7):1909
12. Tang Z, Zhang L, Feng W, Guo B, Liu F, Jia D (2014) Macromolecules 47(24):8663
13. Wei Y, Wang B, Wu J, Yang R, Dunn ML (2012) Nano Lett 13(1):26
14. Smith RF, Eggert JH, Jankowski A, Celliers PM, Edwards MJ, Gupta YM, Asay JR, Collins GW (2007) Phys Rev Lett 98(6):065701
15. Zhang R, Germann T, Wang J, Liu XY, Beyerlein I (2013) Scripta Mater 68(2):114
16. Misra A, Demkowicz M, Zhang X, Hoagland R (2007) JOM 59(9):62
17. Reed EJ, Armstrong MR, Kim KY, Glownia JH (2008) Phys Rev Lett 101(1):014302
18. Desplanque L, Peytavit E, Lampin JF, Lippens D, Mollot F (2003) Appl Phys Lett 83(12):2483
19. Galiotis C, Frank O, Koukaras EN, Sfyris D (2015) Ann Rev Chem Biomol Eng 6:121
20. Gutkin MY, Ovidko I (2006) Appl Phys Lett 88(21):211901
21. Rao C, Cheetham A (2001) J Mater Chem 11(12):2887
22. Min K, Aluru N (2011) Appl Phys Lett 98(1):013113
23. Wang C, Liu Y, Lan L, Tan H (2013) Nanoscale 5(10):4454
24. Sajanlal PR, Sreeprasad TS, Samal AK, Pradeep T (2011) Nano Rev 2(1):5883
25. Roundy D, Krenn C, Cohen ML, Morris J Jr (1999) Phys Rev Lett 82(13):2713
26. Ogata S, Li J, Yip S (2002) Science 298(5594):807
27. Iskandarov AM, Dmitriev SV, Umeno Y (2011) Phys Rev B 84(22):224118
28. Mara N, Bhattacharyya D, Hirth J, Dickerson P, Misra A (2010) Appl Phys Lett 97(2):021909
29. Plimpton S (1995) J Comput Phys 117(1):1
30. Stuart SJ, Tutein AB, Harrison JA (2000) J Chem Phys 112(14):6472
31. Foiles S, Baskes M, Daw MS (1986) Phys Rev B 33(12):7983
32. Huang SP, Mainardi DS, Balbuena PB (2003) Surf Sci 545(3):163
33. Li HY, Ren XB, Guo XY (2007) Chem Phys Lett 437(1–3):108
34. Li J (2003) Model Simul Mater Sci Eng 11(2):173
35. Lee JH, Loya PE, Lou J, Thomas EL (2014) Science 346(6213):1092
36. Ekimov E, Gavriliuk A, Palosz B, Gierlotka S, Dluzewski P, Tatianin E, Kluev Y, Naletov A, Presz A (2000) Appl Phys Lett 77(7):954
37. Cheng Q, Wu H, Wang Y, Wang X (2009) Comput Mater Sci 45(2):419
38. Zhao F, An Q, Li B, Wu H, Goddard W III, Luo S (2013) J Appl Phys 113(6):063516
39. Colombo L, Giordano S (2011) Rep Prog Phys 74(11):116501
40. Luo SN, Han LB, Xie Y, An Q, Zheng L, Xia K (2008) J Appl Phys 103(9):093530
41. Bringa EM, Traiviratana S, Meyers MA (2010) Acta Mater 58(13):4458
42. Giovannetti G, Khomyakov P, Brocks G, Karpan VV, Van den Brink J (2008) Kelly PJ Phys Rev Lett 101(2):026803
43. Park HS, Gall K, Zimmerman JA (2006) J Mech Phys Solids 54(9):1862
44. Belytschko T, Xiao S, Schatz G, Ruoff R (2002) Phys Rev B 65(23):235430
45. Zhao H, Aluru N (2010) J Appl Phys 108(6):064321
46. Zhu T, Li J, Samanta A, Leach A, Gall K (2008) Phys Rev Lett 100(2):025502
47. Chang WJ (2003) Microelectron Eng 65(1–2):239
48. Heino P, Ristolainen E (1999) Nanostructured Mater 11(5):587
49. Stukowski A (2009) Model Simul Mater Sci Eng 18(1):015012
50. Liu F, Ming P, Li J (2007) Phys Rev B 76(6):064120
51. Hong Y, Li L, Zeng XC, Zhang J (2015) Nanoscale 7(14):6286
52. Chen F, Ying J, Wang Y, Du S, Liu Z, Huang Q (2016) Carbon 96:836
53. Zhang H, Guo Z, Gao H, Chang T (2015) Carbon 94:60
54. Chang T, Zhang H, Guo Z, Guo X, Gao H (2015) Phys Rev Lett 114(1):015504
55. Xu Z, Buehler MJ (2010) J Phys Condens Matter 22(48):485301
56. Tschopp M, McDowell D (2007) Appl Phys Lett 90(12):121916
57. Tschopp M, McDowell D (2008) J Mech Phys Solids 56(5):1806

Chapter 7
Summary and Future Work

A summary of this thesis is presented, along with main conclusions on the basic out-of-plane mechanical properties of graphene and mechanical behaviours of graphene composites. The novelty of the research conducted in this thesis is addressed. In addition, remaining issues for the design of graphene and graphene composites are discussed, which may provide some useful guidance for engineering applications in the field of nanomechanics.

7.1 Summary

In this thesis, the static and dynamic out-of-plane mechanical properties of graphene are investigated to address three main issues raised in Chap. 1 using theoretical analysis and numerical simulations, which may highly affect the mechanical properties of graphene composites. The main conclusions in this thesis are summarized below:

- The buckling of graphene is affected by size and chirality. When the compressive length is less than 2 nm, the discontinuous effect leads to an inflexible state along the armchair direction. Otherwise, the buckling is determined by the aspect ratio of graphene sheet α: the preferred buckling directions are the zigzag direction at $\alpha < 1$ and the armchair direction at $1 < \alpha < 3$, respectively; when $\alpha > 3$, the buckling is isotropic.
- For the transverse waves in graphene monolayer, the phase velocity becomes chirality-dependent when frequency f is over 3 THz as a result of the chiral difference in bending stiffnesses. Due to the discontinuous effects, the transverse waves become weaker (down to the noise level) along the armchair and zigzag directions at $f > 10$ THz and $f > 16$ THz, respectively.
- Through the interference of transverse waves in graphene monolayer, high-quality dynamic ripples can be obtained at room temperature, and its electronic properties

are improved. The applicable frequencies for these nanodevices in nanoelectromechanical systems are 1–3 THz.
- The defect probability of graphene under atom bombardment is determined by the impact site, and the physical properties and kinetic energy of incident atoms. As the electronegativity of incident atom increases, the incident energy range for one-step substitution widens, while that for generating single-vacancy narrows.
- Grain boundaries can increase the discontinuous effects in graphene, leading to an anomalous twisting strength of the tilt-armchair-edge graphene nanoribbons (tAGNRs) when the width is less than 4 nm. When the width of GNR is over 4 nm, the discontinuous effects induced by grain boundaries cannot lead to a strengthening effect, but deteriorate the twisting strength of tAGNRs due to the structural instability for larger dimensions.
- The out-of-plane deformation of graphene interfaces in nanolayered composites weakens the in-plane strength of composites, but increases the out-of-plane strength.
- During the propagation of shock waves, the graphene interface can be considered as a "free boundary," while this "free boundary" can remain intact without damage.
- In general, the strength of graphene–metal nanolayered composites increases with decreasing repeat layer spacing. However, when the repeat layer spacing is less than a critical value, structural wrinkles undermine the mechanical performance of composites. Therefore, there exists an optimum range of repeat layer spacing for graphene–metal nanolayered composites.
- Graphene interface can impede the propagation of dislocations. In addition, graphene interface can induce a self-healing effect along the out-of-plane direction for graphene–metal nanolayered composites, which is realized by the in-plane lattice constraint.

7.2 Novelty

The novelty of this thesis is summarized as follows.

- The discontinuous effects on the out-of-plane mechanical behaviors of graphene are revealed, which generally induce chirality and size-dependence in mechanical properties (e.g., the chirality-dependent propagation and the frequency limitation of transverse waves in graphene monolayer). The discontinuous effects play a dominant role when the characteristic size is close to the graphene lattice.
- The effects of defects on the out-of-plane mechanical behaviors of graphene are analyzed. In general, defects decrease the characteristic size of graphene, leading to an increase in the discontinuous effects. Inducing defects into graphene can be beneficial to mechanical and electronic properties simultaneously.
- The interfacial strengthening mechanisms of graphene in nanolayered composites are revealed. Some basic principles are raised for the design of graphene

composites considering such factors as blocking dislocations, self-healing and impeding in-plane buckling.

7.3 Future Work

Although the three important issues raised in Chap. 1 have been studied in this thesis, there are still some significant yet unresolved issues listed below:

- The discontinuous effects on the mechanical properties of three-dimensional nanomaterials have not been studied systematically, despite their importance in engineering applications as applicable structural materials.
- There has not been a unified theory to reveal the strengthening mechanisms of graphene in composites considering the graphene/matrix interfacial interactions, the defects in graphene and the microstructures of matrix, which is important for the design of graphene composites.
- The mechanical behaviors of graphene and graphene composites under extreme loading such as bombardment and radiation have rarely been reported, and bear certain engineering applications such as aerospace devices.

Appendix A
C++ Codes to Model Grain Boundaries in Graphene Monolayer

These codes are developed for modeling graphene with grain boundary at different mismatch angles. The program should be compiled and executed on Linux system:

```
g++ main.cpp -o gb.out
./gb.out
```

The content of 'main.cpp' and related commnets are listed below:

```
#include<iostream>
#include<fstream>
#include<cmath>
#include<iomanip>
#include<stdlib.h>
using namespace std;
const double pi=3.1415926;

// Define the mismatch angle.
const double theta=5.5/2;

// Define the number of atom columns
// related to the defects.
const int NUM=3;

// Define the number of atom columns
// along grain boundary within a period.
// HNUM=11 for theta=5.5
// HNUM=5  for theta=13.2
// HNUM=3  for theta=21.7
const int HNUM=11;

// Define the coordinates of the z-axis of the atom
const double z=0.0;

// Define the number of atom columns
// perpendicular to grain boundary.
const int CPNUM=40;
```

```cpp
// Main function
int main()
{
double k=tan(theta/180*pi);
double y0=1.42/sqrt(3);
double dx=1.42*sqrt(3)*cos(theta/180*pi);
double H5=1.42*(cos(54.0/180.0*pi)+cos(0.1*pi));
double H7=1.42*(cos(5.0/14.0*pi)\
        +cos(pi/14.0)+sin(2.0/7.0*pi));
double H=H5+H7;
double DY=H+13*1.42;
double xj[NUM],xk[NUM];
double yj[NUM],yk[NUM];

ofstream wt("right.xyz");

for(int i=1;i<=HNUM;i++)
 {
 if(i==1)
  {
  for(int j=0;j<NUM;j++)
   {
   xj[j]=j*dx;
   xk[j]=j*dx+0.5*dx;
   yj[j]=k*xj[j];
   yk[j]=k*xk[j]+y0;
   for(int m=0;m<CPNUM;m++)
     {
     wt<<"C "<<setw(16)<<(yj[j]+m*DY)
        <<setw(16)<<xj[j]<<setw(16)<<z<<endl;
     wt<<"C "<<setw(16)<<(yk[j]+m*DY)
        <<setw(16)<<xk[j]<<setw(16)<<z<<endl;
     if(xj[j]!=0)
       wt<<"C "<<setw(16)<<(yj[j]+m*DY)
          <<setw(16)<<-xj[j]<<setw(16)<<z<<endl;
     if(xk[j]!=0)
       wt<<"C "<<setw(16)<<(yk[j]+m*DY)
          <<setw(16)<<-xk[j]<<setw(16)<<z<<endl;
     }
   }
  }
 else if(i==2)
  {
  for(int j=0;j<NUM;j++)
   {
   xj[j]=j*dx+0.5*1.42;
   xk[j]=j*dx+0.5*1.42+0.5*dx;
   yj[j]=k*(xj[j]-0.5*1.42)+H5;
   yk[j]=k*(xk[j]-0.5*1.42)+H5+y0;
   for(int m=0;m<CPNUM;m++)
     {
```

Appendix A: C++ Codes to Model Grain Boundaries in Graphene Monolayer

```cpp
      wt<<"C "<<setw(16)<<(yj[j]+m*DY)
         <<setw(16)<<xj[j]<<setw(16)<<z<<endl;
      wt<<"C "<<setw(16)<<(yk[j]+m*DY)
         <<setw(16)<<xk[j]<<setw(16)<<z<<endl;
      if(xj[j]!=0)
       wt<<"C "<<setw(16)<<(yj[j]+m*DY)
          <<setw(16)<<-xj[j]<<setw(16)<<z<<endl;
      if(xk[j]!=0)
       wt<<"C "<<setw(16)<<(yk[j]+m*DY)
          <<setw(16)<<-xk[j]<<setw(16)<<z<<endl;
     }
    }
  }
else
 {
  for(int j=0;j<NUM;j++)
   {
    xj[j]=j*dx;
    xk[j]=j*dx+0.5*dx;
    yj[j]=k*xj[j]+H+(i%2==1)*((i-3.0)/2.0*3.0)*1.42
         +(i%2==0)*((i-4.0)/2*3+1)*1.42;
    yk[j]=k*xk[j]+H+(i%2==1)*((i-3.0)/2.0*3.0)*1.42
         +(i%2==0)*((i-4.0)/2*3+1)*1.42+pow(-1,i)*y0;
    for(int m=0;m<CPNUM;m++)
     {
      wt<<"C "<<setw(16)<<(yj[j]+m*DY)
         <<setw(16)<<xj[j]<<setw(16)<<z<<endl;
      wt<<"C "<<setw(16)<<(yk[j]+m*DY)
         <<setw(16)<<xk[j]<<setw(16)<<z<<endl;
      if(xj[j]!=0)
        wt<<"C "<<setw(16)<<(yj[j]+m*DY)
           <<setw(16)<<-xj[j]<<setw(16)<<z<<endl;
      if(xk[j]!=0)
        wt<<"C "<<setw(16)<<(yk[j]+m*DY)
           <<setw(16)<<-xk[j]<<setw(16)<<z<<endl;
     }
   }
 }
}

wt.close();
system("echo `cat right.xyz|wc -l` > model.xyz");
system("echo \"Created by XiaoYi Liu.\" >> model.xyz");
system("cat right.xyz >> model.xyz");
system("rm -rf right.xyz");
return 0;
}
```

Appendix B
C++ Codes for Binning Analysis

```cpp
#include<iostream>
#include<fstream>
#include<iomanip>
#include<cstring>
#include<stdlib.h>
#include<cmath>
#include<ctime>
using namespace std;
int main(int argc,char **argv)
{
char CfgName[40];
char X_Dim[10];
int StartCfgNum;
int EndCfgNum;
int IntCfgNum;
int P_ColNum;
int Coor_ColNum;
double Box_xl,Box_xh,Box_yl,Box_yh,Box_zl,Box_zh;
double Box_A_start;
double Box_A_end;
double Box_B_start;
double Box_B_end;
double X_start;
double X_end;
double BinLength;
double TolVol;
double X_factor;
double Time_factor;
double P_factor;

clock_t time_start,time_end,time_pro;
time_start=clock();

ifstream Fin(argv[1]);
char script[40];
while(!Fin.eof())
 {
 Fin>>script;
```

```cpp
    if(strcmp(script,"CfgName")==0)
    {
     Fin>>script>>CfgName;
    }
    if(strcmp(script,"StartCfgNum")==0)
    {
     Fin>>script>>StartCfgNum;
    }
    if(strcmp(script,"EndCfgNum")==0)
    {
     Fin>>script>>EndCfgNum;
    }
    if(strcmp(script,"IntCfgNum")==0)
    {
     Fin>>script>>IntCfgNum;
    }
    if(strcmp(script,"P_ColNum")==0)
    {
     Fin>>script>>P_ColNum;
    }
    if(strcmp(script,"Coor_ColNum")==0)
    {
     Fin>>script>>Coor_ColNum;
    }
    if(strcmp(script,"Box_Size")==0)
    {
     Fin>>script>>Box_xl>>Box_yl>>Box_zl>>script
        >>script>>script>>Box_xh>>Box_yh>>Box_zh>>script;
    }
    if(strcmp(script,"X_Dim")==0)
    {
     Fin>>script>>X_Dim;
    }
    if(strcmp(script,"BinLength")==0)
    {
     Fin>>script>>BinLength;
    }
    if(strcmp(script,"TolVol")==0)
    {
     Fin>>script>>TolVol;
    }
    if(strcmp(script,"X_factor")==0)
    {
     Fin>>script>>X_factor;
    }
    if(strcmp(script,"Time_factor")==0)
    {
     Fin>>script>>Time_factor;
    }
    if(strcmp(script,"P_factor")==0)
    {
     Fin>>script>>P_factor;
    }
   }

if(strcmp(X_Dim,"x")==0||strcmp(X_Dim,"X")==0)
```

Appendix B: C++ Codes for Binning Analysis

```cpp
    {
    Box_A_start=Box_yl;
    Box_A_end=Box_yh;
    Box_B_start=Box_zl;
    Box_B_end=Box_zh;
    X_start=Box_xl;
    X_end=Box_xh;
    }
  else if(strcmp(X_Dim,"y")==0||strcmp(X_Dim,"Y")==0)
    {
    Box_A_start=Box_xl;
    Box_A_end=Box_xh;
    Box_B_start=Box_zl;
    Box_B_end=Box_zh;
    X_start=Box_yl;
    X_end=Box_yh;
    }
  else if(strcmp(X_Dim,"z")==0||strcmp(X_Dim,"Z")==0)
    {
    Box_A_start=Box_xl;
    Box_A_end=Box_xh;
    Box_B_start=Box_yl;
    Box_B_end=Box_yh;
    X_start=Box_zl;
    X_end=Box_zh;
    }
  else
    {
    cout<<"Error: The X dimension is not defined!"<<endl;
    return 0;
    }

  double Shock_Area;
  Shock_Area=(double)(Box_A_end-Box_A_start)\
             *(Box_B_end-Box_B_start);
  double Length;
  Length=X_end-X_start;
  double BinVol;
  BinVol=BinLength*Shock_Area;
  long int BinNum=(long int)(Length/BinLength)+1;

  ofstream XT("x_t.dat");
  ofstream XT_gnu("x_t.gnuplot.dat");
  XT<<setw(20)<<"x"<<setw(20)<<"timestep"
    <<setw(20)<<"Pxx"<<endl;
  XT_gnu<<setw(20)<<"x"<<setw(20)<<"timestep"
        <<setw(20)<<"Pxx"<<endl;
```

```cpp
for(int i=StartCfgNum;i<=EndCfgNum;i=i+IntCfgNum)
 {
 char Incfg[50];
 ofstream tmpf("tmp.name");
 tmpf<<CfgName<<"."<<i<<".cfg"<<endl;
 tmpf.close();
 ifstream tmpr("tmp.name");
 tmpr>>Incfg;
 tmpr.close();
 remove("tmp.name");

 cout<<"Posting "<<Incfg<<" ..."<<endl;

 long int AtomNum;
 int ColNum;
 ifstream read_cfg(Incfg);
 char tmp[40];
 read_cfg>>tmp>>tmp>>tmp>>tmp>>AtomNum;

 while(!strcmp(tmp,"entry_count")==0)
  read_cfg>>tmp;
  read_cfg>>tmp>>ColNum;

 for(int i2=0;i2<(ColNum-3);i2++)
               read_cfg>>tmp>>tmp>>tmp;

 double * P      =new double[BinNum]();
 int     * N     =new int[BinNum]();

 for(long int j=0;j<AtomNum;j++)
  {
  double tmp_d;
  int dc=abs(P_ColNum-Coor_ColNum);
  double x;
  double PerP;
  read_cfg>>tmp>>tmp;

   if(Coor_ColNum<P_ColNum)
    {
     for(int tmp1=0;tmp1<Coor_ColNum-1;tmp1++)
          read_cfg>>tmp_d;
     read_cfg>>x;

     int NP=(int)((x-X_start)/BinLength);
     for(int tmp1=0;tmp1<dc-1;tmp1++)
          read_cfg>>tmp_d;
     read_cfg>>PerP;
     PerP=PerP/BinVol;
     P[NP]=P[NP]+PerP;
     N[NP]=N[NP]+1;
```

Appendix B: C++ Codes for Binning Analysis 101

```
     int LE=ColNum-P_ColNum;
     for(int tmp1=0;tmp1<LE;tmp1++)
      read_cfg>>tmp_d;
     }
   else
     {
     for(int tmp1=0;tmp1<P_ColNum-1;tmp1++)
      read_cfg>>tmp_d;
     read_cfg>>PerP;

     for(int tmp1=0;tmp1<dc-1;tmp1++)
      read_cfg>>tmp_d;
     read_cfg>>x;

     int NP=(int)((x-X_start)/BinLength);
     PerP=PerP/BinVol;
     P[NP]=P[NP]+PerP;
     N[NP]=N[NP]+1;

     int LE=ColNum-Coor_ColNum;
     for(int tmp1=0;tmp1<LE;tmp1++)
      read_cfg>>tmp_d;
     }
    }
  for(int k=0;k<BinNum;k++)
   {
   if(N[k]!=0)
    if((double)(BinVol/N[k])<=TolVol)
     {
     XT<<setw(20)<<(X_start\
         +0.5*(2*k+1)*BinLength)*X_factor
        <<setw(20)<<i*Time_factor<<setw(20)
        <<-P[k]*P_factor<<endl;

     XT_gnu<<setw(20)<<(X_start+0.5*(2*k+1)*BinLength)
             *X_factor<<setw(20)<<i*Time_factor
             <<setw(20)<<-P[k]*P_factor<<endl;
     }
    }
XT_gnu<<endl;
free(P);free(N);
time_pro=clock();
cout<<"Spend time "<<setw(20)
    <<(double)(time_pro-time_start)\
        /CLOCKS_PER_SEC<<" s."<<endl;
cout<<"End Posting "<<Incfg<<" ..."<<endl<<endl;
}

time_end=clock();
```

```
cout<<endl;
cout<<"The total running time is:"<<setw(20)
    <<(double)(time_end-time_start)\
      /CLOCKS_PER_SEC<<" s."<<endl;
XT.close();
XT_gnu.close();
return 0;
}
```

CPSIA information can be obtained
at www.ICGtesting.com
Printed in the USA
LVHW061208250619
622142LV00004BA/74/P